T0130940

The Advanced Materials Revolution

The Advanced Materials Revolution

Technology and Economic Growth in the Age of Globalization

Sanford L. Moskowitz

A John Wiley & Sons, Inc., Publication

Published by John Wiley & Sons, Inc., Hoboken, New Jersey.
Published simultaneously in Canada.

Library of Congress Cataloging-in-Publication Data:

The advanced materials revolution : technology and economic growth in the age of globalization / edited by Sanford L. Moskowitz.
p. cm.
Includes index.
ISBN-978-0-471-61526-2 (cloth)
1. Materials—Research. I. Moskowitz, Sanford L.
TA403.A27 2008
620.1'1—dc 22 2008021432

Printed in the United States of America.

10 9 8 7 6 5 4 3 2 1

To Rose and Becky

Contents

Preface

This book is a product, or confluence, of the three professional paths that I have taken in life: Doctoral student, international management consultant, and academician.

The origins of this book can be traced back to my doctoral dissertation (Columbia University, 1999) and to the projects I undertook as a consultant to government agencies, think tanks, and international corporations since the 1980s. My dissertation dealt with what I refer to as the first advanced-materials revolution; that is, it was a history of global petrochemical innovation—fuels, polymers, and catalytic processes—from 1900 to 1960. This study examined which forces spurred on, and which blocked, technology growth and the economic, technical, and social implications that resulted. From this research, I became acutely aware how critical new materials are to a society and to a nation's economy and ultimate level of competitiveness. My rather extensive work as an economic, market, and technology consultant to government and industry has intensely, and happily, complemented these research interests. As project manager on a variety of assignments, I became, by the mid-1980s, aware that something new and exciting was happening in the world of technology creation as a new generation of materials technology was in the offing. Many were still in development but were most promising, others had already made a major impact on U.S. and global economies, and still others were just beginning to take their first steps into the marketplace. Over the years, I tracked many of these technologies as they evolved from the laboratory to initial market entrance to diffusion within the economy.

By the 1990s, I came to understand that these new material technologies were a decidedly different animal from the earlier petrochemical innovations that I had researched and wrote about. It was clear to me that the era of the petroleum-based "supermolecule" was over in the sense that the radical new developments now emerged from the manipulation of the internal world of individual molecules, rather than the mega-linked-chain structures that characterize nylon, synthetic rubber, polyethylene, and the other polymers with which we are all now familiar. Moreover, with the passing of these materials from the high-technology limelight, I also noticed how often smaller firms, many of them no more than start-up operations, took the place of the traditional mega-corporations in developing the most important new material innovations. In so many cases, these smaller firms licensed patents held by universities. It seemed no coincidence to me that by the 1990s, many U.S. research universities had established various forms of technology transfer offices and incubator facilities to guide the transfer of their faculty's research efforts to the outside world. And increasingly, the large firm, no longer undertaking brand new technology, came calling on the smaller company for ideas, patents, and licenses, often times acquiring it to possess its valuable intellectual property. But this was not all. In my consulting work (and some initial research into this phenomenon of the "new" advanced materials), I began to see the rise of a number of high-technology industrial "clusters" within the United States that seemed to crystallize around a firm or two involved, either directly or indirectly, in these new materials. This meant a trend away from centralization (away from large firms and one dominant cluster, i.e., Silicon Valley). At the least, these former start-ups were early entrants into the cluster, grew along with it, and became very quickly leading members of the group. Nor did these clusters arise automatically, but seemed to cohere and expand through the agency of strong-willed, multitalented individuals who were known and respected by, and easily moved within circles of, different disciplines and fields: scientific, technical, market, financial, regulatory, and political.

These strands of thinking emerged over time as I talked with various individuals and researched different technologies and markets, both here and abroad, as both inquisitive academic and project-driven managing consultant. As I probed deeper into this world of advanced materials, I began formulating some interesting questions. Could we be witnessing a new type of paradigm for technology change? Can we properly claim that advanced materials is "the" central technology, the "straw the stirs the drink," of technological change in the late twentieth and twenty-first centuries? If so, has the small or medium-sized advanced-material firm taken over for the large multinational chemical company the role of pioneer in the technology creation game? Has the university's materials engineering departments, through its technology transfer offices and associated incubators, taken on a more important role in technology creation than it has ever had in the past? From this, can we say that technology creation today is far more a grass-roots, "from-the-bottom" process than the far more familiar "top-down" (large corporations, Federal government) mechanism that has brought the world atomic energy, synthetic rubber, and the mega-technologies of the past? If this is the case, then can we be seeing the rise of a "seamless web" model that links the original academic research with start-up technology and then entrance into the larger economy?

It was at this time that John Wiley contacted me to write an article on advanced materials as an entry for its well-known *Kirk-Othmer Encyclopedia of Chemical Technology*. This project helped to crystallize in my mind the above questions and issues. It also brought me to an understanding of how these newest technologies are so central to a country's economy and, ultimately, its competitiveness in the world.

As a professor of international business (at St. John's University, Collegeville, Minnesota), I understood that the rise of these advanced materials is taking place at the same time when the world is becoming more globalized. From my most recent research on globalization, technology, and economic growth, I began to conceive of a larger work, an expansion of my article, that would employ advanced materials as a superb prism through which we might understand the competitive nature of countries and regions in the twenty-first century. This book is the result.

This book is a true hybrid. It is certainly part history, albeit of a more recent vintage, and as such it is a narrative of what the book calls the advanced-materials industry. I discuss the origins of advanced materials in the 1970s and trace their development to the present time. I also link the current crop of advanced materials to the earlier period of petrochemical polymers in order to better understand the similarities and, even more importantly, the distinctions between the two materials revolutions. This book is also descriptive in that it is part technology assessment, industry analysis, and market and product forecast. It describes in some detail what the new materials are and the nature of their markets. It discusses the promising technologies, companies, and regions within the United States and internationally that are on the verge of reaching their full bloom, and others that are not, for whatever reason, able to move forward. These parts of the book should be of interest to investors, entrepreneurs, companies, consulting firms, and universities that want to take their internal research into the real world of markets and competition.

In addition to it being a historical analysis and industry/market description, it is also a thematic analysis on issues that are front and center in international business practice and studies. In this role, advanced materials serve as a very useful guide in helping us understand the forces, institutions, and actors involved that determine national and regional competitive advantage in the globalizing world today. There is a growing division between the United States and the European Union (EU) with respect to productivity and economic performance. Because the EU has grown in geographical extent and resources and has become a more integrated market, it is believed it should be an increasingly worthy competitor to the U.S. economic hegemony, but in fact we see this is not the case. That is, we see a "divergence" rather than "convergence" in the relative performance of the United States and the EU over the last decade and a half. But this is certainly not the "flatness" we ought to be seeing in a growing globalized world, especially in the two regions that are now so apparently equal in their access to important factors of intellectual and material factors of production.

Certainly, many factors account for observed differences in productivity and economic growth: cultural issues (e.g., the greater "leisure culture" of the Europeans), macroeconomic trends (e.g., deficits and currency fluctuations), and sociopolitical movements (demographic shifts and conflicts) and so forth. This book does not mean to dismiss these important forces. However, the evidence—quantitative and qualitative—all points to a fundamental divergence between the United States and the EU stemming from technological differences. Simply put, the United States has been more successful than the EU in recent years in creating and applying the newest technologies; more than ever before in history, technological change is related to economic development and growth; and the evidence shows a growing importance of advanced materials as a component of total technological expansion. That is to say, now more then ever, a country's technological capability closely shadows its creativity in developing and harnessing the new "advanced" materials. This issue of global technology, as embodied in advanced materials, international competitiveness, and the notion of convergence versus divergence is of great interest today among international business students, scholars, and executives. This book then will be useful to this audience as well as those involved in the study and practice of the management of technology and global development. The model the book proposes for competitiveness not only advances this discussion of the United States versus the EU, but also helps better understand whether Asia is, as generally believed, set to become the economic power of the twenty-first century. We discuss this in the concluding chapter from the point of view of advanced materials, clusters, and competitiveness.

A question I had to address early on was whether there is, in fact, a coherent set of products and processes that can be placed in an advanced-material category. They certainly are a diverse, wide-ranging set of technologies. I have, in fact, relied on what industry specialists and entrepreneurs from the United States, Europe, and Asia have told me when asked what they would include in such a category. Although no two lists were exactly the same, it was clear that certain entries were common to many responders. I have called these the advanced-material category. This category includes not only products themselves (e.g., nanotubes) but processes and instruments used to make and monitor the materials. As for whether we are talking about an industry, it is clear that we see certain common characteristics of this group, such as the importance of small- and medium-sized enterprises (SMEs), so that we can discuss the industry in terms of a "seamless web" structure. It is also apparent that this group of products is linked together by common process technology in a manner similar to the petrochemical industry.

In this sense then, such as the common theme that the new materials are made by intramolecular manipulation rather than creating large molecular chains, we are dealing with a coherent set of products that constitute an industry. Although nanotechnology is part of the story of advanced materials, and a rather large part at that, it is not the whole story by any means. A number of advanced materials, such as advanced alloys, organic polymer electronic materials, and biorefinery products, are not part of the nanotechnology firmament. In fact, many of the advanced materials that have already had a major economic impact cannot be called nanomaterials. Therefore, the book distinguishes between "nanotechnology" and "advanced materials," while still recognizing the real and important link between the two realms of technology.

This book is based on a wide variety of sources. These include government (EU and U.S.) reports and studies; industry studies; articles in international business and trade journals; interviews conducted with entrepreneurs, executives, and academics in the field; scholarly reports and articles; and global company and organization websites. I have incorporated as well (1) the database I have collected over the years on doing work in the advanced-material field, including product and industry studies I have undertaken as well as discussions and interviews I have conducted with specialists in the advanced-material field; and (2) data, information, and insights I have obtained in teaching international business courses over the last several years. Guidance and insights from participants at international conferences at which I presented research papers in the area of global technology and international competitiveness have been invaluable.

Another useful source has been *Nanoinvestor News* (www.nanoinvestornews.com). Especially helpful have been the industry and corporate profiles. While focused on nanotechnology, these profiles include many advanced-material firms within the United States and internationally. The site was very helpful in guiding me to many of the most important new materials and their firms globally. Degree of "importance" could be determined by how many articles were written about the firm, which was indicated at the end of each profile. Because the site encompassed the world's advanced technology firms, it minimized the book being biased towards U.S. firms. I was able to fine tune the list through discussions with colleagues, especially when attending conferences in Europe, and in articles in business and trade journals. I also incorporate into this book an unpublished study on advanced materials that I undertook for the state of Virginia. This short examination of the "industry" contains useful information and analysis that has relevance to the present work.

I constructed a number of tables and charts from these sources. Many of these tables are original and specific to U.S. and international advanced materials. Unless otherwise specified, these tables, including those forecast past 2007, have not been previously published. I have developed these tables to conform as closely as possible to the evidence I have at hand.

I have included a number of citations and endnotes after each chapter. I have in no way attempted to drown the reader in citations, as might be expected in a purely academic work. I have however placed citations at points in the discussion that I felt would be useful to the general reader and which indicated what sources I used to reach certain conclusions or make specific claims. At various points, I listed all citations used at the beginning of a section. In other cases, such as direct quotes, I generally included a specific citation and page number. In those cases when page numbers were not available (such as some online documents), I simply cited the document as a whole.

Finally, this book is not meant to be the final word on this subject. The advanced-material industry is too dynamic and far reaching to be encompassed in a single work. The discussions, conclusions, and implications presented in the following pages are based solely on my own experience, research, and insights in the field. The book is actually the first to tackle advanced materials as a global industry with international reach and impact. The book is also not meant to be a paean to American talent and economic might. If the United States has held out longer as a technological leader than generally believed, even in the face of globalization, its position at the top remains quite shaky The fact that it is, at this writing, in the throes of economic uncertainty and that the rise of Asia as a competititve power in the world is clearly in evidence ought to give even the most inveterate supporters of "The American Century" serious and troubling pause. The question, then, is whether the strengths I describe as wielded by the United States in technology creation and economic growth are gaining additional power for an even brighter future, or quickly losing their once-considerable influence around the world. If the latter is the case, does this then mean a surging European or Asian presence in advanced technology markets, as one or both rush in to fill the vacuum created and thus rise to, or even supersede, the competitive level of the United States? This book hopefully will add to the dialogue and suggest possible directions for future and useful studies on the role of advanced materials as the global technology of the twenty-first century.

SANFORD L. MOSKOWITZ

Saint Joseph, Minnesota
October 2008

Acknowledgments

Various parts of this book benefitted in significant ways from the expert advice and suggestions provided by a number of persons. I am most indebted to Professor Alan Brinkley and Professor Walter Metzger who helped guide me through my doctoral dissertation (Columbia, 1999), portions of which I incorporated into critical parts of this book. I have also received continual advice and guidance that has proven of immense value to me from Professor Terry Reynolds (Michigan Technological University), especially in those parts of the book dealing with the innovation process and themes related to the history and social context of technology. I am indebted as well to Professor Stephen Stumpf of Villanova University for his generous support and guidance in the early phases of this project. I am grateful to Professor Jonathan Doh (Villanova University) for introducing me to the intricacies of United States–European Union trade disputes—particularly as related to the "genetically modified food" issue—and his guidance in the area of managing change within the "multienvironmental" context; this book benefitted greatly from his writings and insights on these issues.

This book incorporated in a significant way my experience as a consultant to the advanced technology sector in Washington DC in the late 1980s and 1990s. Richard Cooperman and Janice Lipson were my invaluable guides and advisors in this work. They proved especially important in sharpening my skills in market analysis and forecasting and in understanding the legal and regulatory mechanisms of high-technology industries.

I am most grateful to my colleagues in the Departments of Management and Economics at St. John's University and the College of St. Benedict (Collegeville, Minnesota) for their help and advice during the final stages of this book. I want to thank Professor John Hasselberg and Professor Joseph Friedrich for their unstinting support of my research and in their efforts on my behalf to obtain funding to present my findings at international conferences in the United States and Europe. The suggestions and guidance that I received from international business scholars at these conferences proved very useful in completing the final chapters of the book. I am indebted as well to Professor Louis Johnston for his excellent insights into the economic history of innovation that, both directly and indirectly, entered materially into this book. Of particular relevance to this study were Professor Johnston's discussions with me regarding the relationship over time between technology and productivity and the relevance of the recent work done by economic geographers on the "convergence" versus "divergence" issue.

I have been the grateful beneficiary of the talents of superb research assistants that have aided me significantly in this effort. I am indebted to Mr. Andrew Bruskin who proved to be a first class researcher and analyst. Mr. Bruskin helped me work through some of the more difficult research problems that came my way in the course of this project. He was particularly critical in the areas involving the legal and political issues related to advanced technology and was a very useful guide into state and local mechanisms for high-technology cluster development. This book benefitted greatly from the research assistance of Ms. Roxanne Rabe and Ms. Elizabeth Sturlaugson. Over the course of two summers, they worked tirelessly to help prepare portions of the manuscript for presentation at conferences in Berlin, Paris, Florence, Athens, and Vienna. They aided me as well as copresenters at these conferences and in incorporating the critical input from the conferences into the final manuscript. Their advice and insights in this undertaking helped shape the final results and observations included in the book. I am also indebted to the very excellent work of Judy Shank and Sue Zimmer for their careful editing and efficient preparation of this manuscript.

In the final analysis, of course, I can blame none of the shortcomings of this volume on any of the superb people who have so graciously assisted me on this book. That responsibility I must carry on my shoulders alone.

S. L. M.

Part One

Advanced Materials, Past and Present

The end of the last century ushered in a revolution in technology that is still unfolding. The emergence of the advanced materials industry, beginning in the early 1980s, ushered in one of the most dynamic and important chapters in U.S. and international industrial history. These revolutionary materials possess new and different types of internal structures and exhibit novel physical and chemical properties with an unprecedented range of application. They have already gained a strategic foothold in international economies. They continue to diffuse into and transform the world that we know, and the society we will come to know over the next century and beyond. By 2020, they will generate direct sales worldwide of hundreds of millions of dollars. These materials invade and restructure virtually all the major industrial sectors. They particularly impact the computer and information sector, redefine the nature of energy creation and transmission, and are leading change of epic proportions in the biomedical, healthcare, transportation, and manufacturing industries. The very nature and trajectory of twenty-first century technological change, and the productivity growth and economic progress that follow in their wake, fundamentally hinges on these essential building blocks of modern life.

These materials include the new generation of metals, advanced plastics and ceramics, and biosynthetics. Beyond these metals and synthetic organic materials, the advanced materials field finds itself embedded within the very heart of the emerging world of nanotechnology. Indeed, the so-called nanomaterials, more than any other area of nanotechnology, amply testifies to the commercial possibilities of this new world of the very small. The techniques, instruments, and knowledge of nanotechnology open up the vast possibility of the manipulation and restructuring of molecular units within many substances and material systems. It is in this realm that some of the most exciting and economically important developments emerge, including nanotubes and nanospheres, thin films, nanofibers, and nanocomposites. The alteration of the vast spectrum of the world's materials is the most important application of nanotechnology as a whole. Nanotechnology is the key in the coming new generation of polymers, cutting tools, coatings, optical components, catalysts, corrosion-resistant materials, and drug delivery systems.

Trends in the patenting in the field of nanotechnology clearly reflect the important role of new materials in the field, at least in terms of research interest and applications. An influential report put out in 2003 by Lux Capital shows the rapid rise in attention being paid to the field of nanotechnology. From 1995 to 2003, the number of articles published by Dow Jones Publications that mention the word "nanotechnology" increased from less than 20 to over 2,000. If we consider all U.S. business and technical publications, then this range of increase was from under 50 to nearly 5,000, with this latter figure accounting for approximately 85% of all U.S. business and technical publications [1]. Moreover, the major commercial interest in nanotechnology is in the new materials (the so-called nanomaterials). The *Journal of Nanotechnology Research,* which classified the total number of patents issued in nanotechnology from 1976 to 2002, according to specialty, shows that the two most important groups in terms of patenting activity resided in the new materials arena: "Drugs" and "Chemistry: Molecular Biology." If we also include the fields of "Or-

TABLE I.1. Patent trends: nanotechnology and new materials

Field name	Number of Patents (1976–2002)
Drug: bio-affecting and body-treating compositions	10,866
Chemistry: molecular biology and microbiology	7,946
Radiant energy	4,657
Stock material	3,939
Solid-state devices	3,933
Semiconductor device manufacturing: process	3,877
Organic compounds	3,756
Chemistry: natural resins, derivatives	3,753
Optics: systems and elements	3,404
Coating processes	3,265
Chemistry: analytical and immunological testing	3,027
Radiation imagery chemistry: process, compositions, and products thereof	2,983
Optics: measuring and testing	2,957
Information storage and retrieval	2,310
Electrical nonlinear devices	2,286
Chemistry: electrical and wave energy	1,864
Chemical apparatus and process disinfecting	1,829
Coherent light generators	1,775
Compositions	1,680
Multiplex communications	1,638
Total	71,745

Source: [2].

ganic Compounds," "Chemistry: Natural Resins," and "Coating Processes," we see that the materials field is in the top ten (out of 20 total fields) in terms of patenting. Overall, more than one-half of all these patents (53%) involved research into new nanomaterials.

Based on the above, the patenting of nanomaterials in one form or another occupies three-quarters of nanotechnology's commercial development. This percentage is likely to grow over time as industry and consumers demand more, cheaper, and different materials.

The current advanced materials revolution, which began in the 1980s, is potentially the most significant and far reaching technological movement since the nineteenth century, not just economically but socially and culturally as well. Although it evolved out of the earlier technologies, it is decidedly not a simple extension of those prior achievements. It has pursued it own, unique trajectory.

One of the major differences between earlier periods of the super molecule and today is that the new-materials revolution is taking place within the context of the globalization movement. This is of critical importance in terms of the distribution of wealth creation internationally. As we shall see, increasingly, the creation of new materials, because of their centrality in industry and society, has become a leading driver of economic growth of nations. Since the 1980s, technological change and economic progress have grown more mutually interdependent with both of these closely shadowing new-materials development. Accordingly, an examination of global technology, and advanced materials in particular, helps determine which nations and regions are gaining and will gain, and which are losing and will lose competitive advantage in the world economic system.

This question lies at the very heart of a fundamental dispute in international business today. One side of the debate sees a steady spreading of economic benefits to more and more countries worldwide, with a resulting social enlightenment following in its wake. But is this the case? Are we indeed seeing the leveling effects of globalization? If, as Thomas Friedman believes, globalization has rendered the world "flat," does this mean that the United States must relinquish its role as main competitor internationally to other regions and countries, such as Europe or Asia? [3] Much has been written recently on the decline of the United States as an economic power and the surging competitiveness of erstwhile backwater economies, especially within the European and Asian countries. Thus, Thomas Friedman recently wrote: ". . . the biggest challenge . . . facing us today [is] the flattening of the global economic playing field in a way that is allowing more people from more places to compete [with the United States] [4].

There is no question that the United States faces greater competition internationally than ever before. But can we say that the United States has been losing ground in this new "flattened" world? One factor that is called into play by those who see globalization as a losing proposition for the United States is the declining academic performance of American high school and college students relative to those of the rest of the world. If the United States continues to lead the world in the quality of its graduate schools, students from Europe and Asia are the ones, some argue, taking maximum advantage of these institutions and then, increasingly, taking the knowledge and skills they have learned and transferring them back to the benefit of their home countries. Greater opportunities in, and enticements offered by governments of, formerly less developed countries, as well as ease of long-distance transportation and the rise of the Internet, entice these superior talents, nurtured by the American higher-educational system, back to their homelands.

And what of Europe, said to be reborn as an economic power? As the European Union expands and becomes more integrated, it gains a number of important advantages. The European Union benefits from an expanding area and population, single market and economies of scale, single currency (and, therefore, cheaper capital and more competition), and a federal government increasingly dedicated to becoming competitive with the United States through centralized policies. More specifically, the European Union increasingly coordinates funding and development activities in the most advanced scientific areas. We might expect then that, given these developments, the European Union should have begun to compete well with the United States, certainly beginning in the 1990s when integration proceeded apace.

Countering this "convergence" view of globalization is the "divergence" or "uneven distribution" model by which economic activity and wealth are concentrated in certain countries while the less developed nations lose out in the globalization game. Supporters of this model note that the United States itself has been benefiting mightily from the globalization movement, in fact retaining control, either directly or indirectly, over a preponderance of the world's wealth. In this scenario, the United States then remains competitive because of the support of government and America's immense scientific, technical, and financial resources.

The question we are raising here, that is, whether, in this global world, economic activity is unevenly distributed and concentrated in discrete locations, say the United States or the European Union, or whether it is becoming more widely dispersed and "evened out" in a continuous pattern of growth, is a critical one. This issue relates not only to what have been the trends in economic activity geographically, but what patterns of economic activity we can expect over the next decade and beyond. Is modern technology today, which without question propels economic growth, a benefit to the economic development of all nations through the globalization process? Or are the fruits of a home-grown capability not as easily transferred to other countries, as is often assumed by those who preach the gospel of a benevolent, democratic globalization? It is certain that the temporal and geographical evolution of the new-materials revolution is playing, and will continue to play, a central role in this debate. The limits to which one nation or region can outpace other countries and regions depends more desperately than ever before in history on how successful it is in embracing, commercializing and applying the fruit of these new technologies.

This story concerns itself with two primary regional actors on the global stage: the United States and the European Union. We witness as these two leviathan states go head to head in a contest to determine which holds today, and will hold, the greater competitive advantage. Our time frame then is the near past, the present, and a future projected out to 2030. The approach of comparing and contrasting the United States and the European Union offers significant advantages. As the European

Union has enlarged itself and pushed for greater cohesion through market integration, it has become the most likely regional candidate, even more so than Asia, to compete against the United States in world markets. The bulk of advanced materials science and technology occurs in these two regions. Indeed, both the United States and the European Union governments have undertaken central policies to advance progress in new-materials technology. American and European firms are most active in establishing joint ventures with one another in researching, developing, and commercializing new material products and processes. These studies then will alert companies, investors, and governments as to the most important material technologies from the United States and the European Union coming on line, the firms developing them, what regions and countries within the United States and the European Union look most promising for advanced material development, what these technologies will be, their economic importance, and potential roadblocks that need watching.

Beyond these particular trends and profiles, we want to better understand the dynamics of regional growth and competitiveness. By all accounts, the European Union, through its evolving and manifest competitive unity, ought to be proceeding strongly as a competitive power. And in fact, in this first decade of the new century, the world community of industrialized nations waxes optimistic regarding Europe's economic future, especially in the wake of trade liberalization and closer economic integration. All this was supposed to make the European Union a mighty competitor to the United States. Recent works by John McCormick (*The European Superpower*) and T. R. Reid (*The United States of Europe: The New Superpower and the End of American Supremacy*), for example, alert us in unmistakable terms to the putative emergence of Europe as the preeminent force on the world's economic (and political) stage. These books remind us that, even if the Europeans cannot match the sheer might of the U.S. military machine, they will prevail as the next superpower by wielding their continent-wide economic and political power to dominate global market shares in the most important, technology-driven industries [5].

This study explores this question to better understand the competitive position of modern-day Europe relative to the United States. Since technological change is so central to economic growth, productivity, and, therefore, competitiveness, we use America's and Europe's position in the advanced materials technologies as bellwether and proxy for their different abilities to maintain and exploit regional competitive advantage in the twenty-first century.

At the end of the day, it is hoped that, by examining the shifting competitive positions of these two seminal powers through the prism of advanced material technology, we will shed a penetrating light on the continuing debate on the nature of globalization and its impact on economic growth across geographic regions. In doing this, we want to see whether the economic fruits of the new and emerging technologies are being spread evenly or whether they congregate and amass in one region to the detriment of others. If the latter is true, why is this divergence the case when the general belief is one of leveling through "openness," transparency, and striving for common international standards that supposedly lies at the very core of globalization? And finally, what sort of model can we construct that allows us to better understand the wealth formation within developing countries and regions, such as in Asia, and predict the competitive position of China and India during the first half of the twenty-first century.

The important part of this story is the basket of new and emerging advanced material technologies that have been coming into commercial existence since the mid-1980s, for these lie at the very heart of modern economic progress. We refer to them collectively here as the "advanced materials industry." The petrochemicals sector produces many different materials but is a recognized industry that uses common process technology [6]. So too does the new generation of advanced materials. In the case of the petrochemicals group of the 1950s and 1960s, for instance, process innovation built up macromolecules from smaller units. The advanced materials of the twenty-first century employ processes to reconfigure separate atoms and atom groups within discrete molecular units. This ultimate and shared goal over a riot of new and diverse substances is the "red line" that links this assemblage of materials into an integrated and consistent industrial unit.

This book is structured to examine these products and processes, this advanced materials industry, in detail: what these materials and processes are; why they are important for society now and as we proceed into the twenty-first century; what pressing themes and issues of technology development, globalization, and the distribution of economic activity that this industry brings to the fore and forces us to address; and why and in what way all this is important for us to know as we try to sort out the rise and fall of economies as the world becomes more complexly integrated and dependent on the new technologies.

In Part I, we introduce the reader to the "new" materials of the past and the current generation of advanced materials. We attempt to understand what exactly these new technologies are and how they evolved and diverged from past industrial revolutions (Chapter 1). In Part II, we dissect the great opportunities—the potential—of these materials as an economic force and the very real risks involved that cannot be ignored and that can scuttle even the most trenchant hopes and projections of the materials' economic impact. We first take up the potential opportunities. We explore the current and anticipated application of these materials within society. By understanding these technologies in more detail and their critical place within society, we sharpen our appreciation for the massive impact these technologies are having, and might increasingly have, on national economies. Then, using past successful high-technology products as models, we develop optimal (best case or "upper limit") market demand projections for these technologies globally and for the United States and Europe separately (Chapter 2). These projections take a highly optimistic view of the world today in that they assume the "convergent" model of globalization. We describe the reasons why convergence ought to occur and discuss the positive implications that flow directly from this assumption. This means that we allow that the European Union will, over time, increase the share of the wealth that it will capture from new and emerging technologies until it reaches a level more or less equal to that of the United States. But this is essentially an ideal construct. We must then come down to earth, upon which the reality exists that there are forces that can hinder, and even derail, the best and brightest of intentions. We then ask what factors, in reality, may emerge to thwart this picture of an equally beneficent globalization. We begin doing so by looking in some detail at what exactly is the relationship between the new and emerging technologies and economic growth, and, ultimately, the degree of competitiveness (Chapter 3). It is here that we begin to see economic problems facing the European Union, even as it tries mightily to measure up to the United States in terms of productivity and economic growth. We then go on to discuss the central position of advanced technology, driven by the new and emerging materials, in modern economic performance, especially within the information technology (IT), energy, and biomedical fields and, finally, show the sluggish performance of Europe in new technology growth vis-à-vis the United States.

At this juncture, we begin to understand that the convergence model of globalization must be seriously questioned. That is to say, if a large, highly industrialized region that is so closely tied to U.S. business and technology such as the European Union is itself not capable of meeting its competitor on equal terms, if we do not see a leveling effect even in this case, what chances can there be that less developed regions and countries will do so in the foreseeable future? We then begin to attempt to understand why the European Union, despite the force of globalization and the extensive and intricate links established between American and European business, has not been able to keep up with U.S. progress and growth. Our discussion then turns to a better understanding of the risks—technical, economic, managerial, and political—that directly face new-materials development and threaten the harnessing of this new and critical technology to economic growth (Chapter 4). In doing this, we can begin to pinpoint where and in what ways countries and regions (such as the European Union) that falter as competitors may be actually increasing the risks of failure while more successful countries and regions (such as the United States) more effectively reduce to a bare minimum, and even transform into an advantage, these very same risks.

This comparative analysis of risk management leads us naturally to understanding how and why countries and regions vary in the ways in which they deal with the risks involved in creating, nurturing, and diffusing essential technologies. Part III allows us to peer into what is arguably the most essential stage in the life of a new technology—research and development (R&D)—for it here here that new technology first sees useful life and becomes prepared for the marketplace. In this narrative, we develop the idea that, whereas all countries (and firms) face these risks, not all countries (and firms) perceive of or handle them in the same way. We compare and contrast R&D strategies in the United States (Chapter 5), and the European Union (Chapter 6). In this part of our story, we will see that, whereas the U.S. firms most essential to advanced materials conduct research, development, and commercialization in ways that reduce these risks, European countries and firms developed R&D strategies that actually play into, nourish, and ultimately heighten these risks. As the Europeans do this, they find themselves falling behind the Americans in successfully creating new materials for the market, even as they continue to excel in exploring the fundamental science.

We then need to place these differing styles of advanced materials creation within their broader technical and social contexts. This affords us a greater, and essential, perspective on technology creation and growth within the advanced materials sector as it takes place globally within the late twentieth and twenty-first centuries. This portion of our story reveals the fundamentals of how high technology commercialization and market entrance must be organized in the twenty-first century in order to reduce the most important risks that can undermine the root competitiveness of a country or region. To this end, Part IV introduces the concept of the "seamless web" in advanced technology development. We discuss the close interlinkages that must exist between the major players of advanced materials technology, including corporations, universities, incubators, and start-ups (Chapter 7). We note in particular how the United States has developed an intricate but highly coordinated network or web that works efficiently to generate new advanced materials technology and thrust these products into marketplace application. We observe how universities and technology-transfer organizations work with closely allied incubators to guide newly born technology from the pristine environment of academia into the start-up firm, the last stop before commercial entrance. In the following chapter (Chapter 8), we describe how, within this seamless web of codependency, mechanisms exist that serve to select the most promising technologies to develop, and weed or filter out those that cannot make the grade. The central role of venture capital dominates these pages. For both chapters, we emphasize the difficulty Europe has been having with developing in any robust way these various actors and with linking these in any coherent and integrative manner into the requisite seamless web structures.

In Part V, we are concerned with how these structures actually organize themselves and function as coherent systems. It is argued that the existence of these types of networks does not mean that new innovation, and the economic growth that follows it, can occur anyplace in society. Rather, these webs of creation and economic progress take root in specific places and for particular reasons. We discuss the nature of these advanced material clusters, the varieties that exist within American society, and the complex dynamics of their evolution and growth (Chapter 9). We then compare the creative thrust of U.S. clusters with the less innovative European cluster models, and explain the underlying cause for these differences. The penultimate chapter (Chapter 10) introduces the fundamental organizing force of America's most innovative advanced materials clusters: the "gatekeeper." We discuss the great variety of gatekeepers that shape and energize creative clusters and we bring to the fore the salient differences that exist between American and European gatekeeping cultures and how these distinctions go to the heart of the "great divide" between the two great powers of this still evolving global era. The concluding chapter (Chapter 11) considers what our foray into global advanced materials tells us about the nature of globalization as it is unfolding today, and how it might proceed in the coming decades. These final thoughts bring into the discussion how what we have learned in the previous pages can help the West better understand and assess the forces that control the competitive power of an emerging Asia.

REFERENCES

1. *The Nanotech Report* (2003), New York: Lux Capital, p. 23.
2. Huang, Z., Chen, H., Yip, A., Ng., G., Guo, F., Chen, Z. K., and Roco, M. C. (2003), "Longitudinal Patent Analysis for Nanoscale Science and Engineering: Country, Institution and Technology Field," *Journal of Nanoparticle Research,* Vol. 5, Nos. 3–4, p. 15.
3. Friedman, T. L. (2005), *The World Is Flat: A Brief History of the Twenty-First Century,* New York: Farrar, Straus and Giroux.
4. Friedman, T. L. (2005), "What, Me Worry?" *New York Times,* April 29, p. 28.
5. McCormick, J. (2006), *The European Superpower,* New York: Palgrave Macmillan); Reid, T.R. (2004), *The United States of Europe: The New Superpower and the End of American Supremacy,* New York: Penguin Press.
6. Trescott, M. (1981), *The Rise of the American Electrochemicals Industry, 1880–1910,* Westport, Connecticut, Greenwood Press; Spitz, P. H. (1989), *Petrochemicals: The Rise of an Industry,* New York: Wiley.

Chapter 1

The Coming of the Advanced-Materials Revolution

In this chapter, we examine in greater detail what the new advanced materials are and where they came from. We arrive at this goal most effectively by providing an historical context out of which this new technological revolution emerged.

Discussions surrounding advanced materials in the literature are fond of stressing a rupture with the past with respect to the science, technology, organizations, and even markets involved. Although important distinctions exist between what we will call this new-materials revolution and previous periods of technological change, these differences should not be given too much emphasis. Indeed, the new-materials revolution descended from that same line of scientific, technical, organizational, and economic development that began in earnest, and is deeply rooted, in the nineteenth century. One cannot understand present developments without reference to these earlier revolutions.

It is fair to say, as one author has written, that "the hallmark of progress in every age has been the way 'materials engineers' worked to improve the usefulness of materials, whether extracting coal or iron ore from the earth or creating new materials from combinations, such as iron and carbon to produce steel." [1] Certainly, previous periods saw the introduction of new materials into the world, from iron and bronze in ancient times to aluminum, stainless steel, gasoline, and synthetic chemicals and resins after 1900. From the late nineteenth century to the late 1970s, there were two distinct periods of new-materials development. The first period, lasting roughly from 1880 to 1930, showcased the famous coal-tar products (in Germany) and the mass manufacture of metals, notably steel and aluminum (in the United States). From the 1930s to the post-World War II decades and up to the 1960s, coal as a raw material for organics and metals gave way to other fossil fuels, in particular petroleum and natural gas. This period witnessed the innovation and mass manufacture of completely new and highly complex man-made petrochemicals. This second materials revolution brought the world advanced fuels and the macromolecular synthetics, including man-made fibers, plastics, and resins. During this period, the United States finally gained technical and economic hegemony over Europe, and especially Germany, in the development, production, and application of advanced materials.

The materials that came to commercial prominence during these two technological revolutions represent more than just technical and commercial accomplishments in their own right, although they were certainly that. The more important ones played pivotal roles in the expansion of other interrelated industries. This technological interdependence proved to be a central process in the rise and growth of economically vital industries. In the late nineteenth and early twentieth centuries, the new metals, such as steel and aluminum alloys, supplied crucial inputs into America's expanding railroad system (revolutionizing the design and construction of both rail and rolling stock), in construction (especially in the building of skyscrapers, bridges, and highways), and in manufacturing (in the construction of new factory buildings, the fabrication of machine tools and equipment, and the design and production of products made within these plants and with these tools and equipment).

In the second period, the new synthetics revolutionized, as well as disrupted, the industrial landscape, first within the United States and then internationally. The large chemical companies, including Union Carbide and Dow, at first led the way to these new petrochemical materials. Then the major refiners, notably Exxon (Jersey Standard), Phillips, and Shell, integrated forward into chemical intermediates and final synthetic products, and became leading innovators in the field. In the postwar period—from the 1940s through the 1960s—as production methods improved and reduced the unit costs of making the new petrochemicals, these materials diffused into and fundamentally transformed those industries that define the contours of what we know of as our modern economy, including semiconductors, communications, energy, transportation, and, increasingly, the Internet and biotechnology.

These two phases of new material development traversed most of the nineteenth and twentieth centuries. This means that they remain intricately linked to what we know as the industrial revolution, especially within the United States. Table 1.1 summarizes this forward progress of innovation during these centuries. The table shows the most important innovations. These may be new materials altogether or, in some cases, new processes to make commercially known materials. The table indicates the year of and country most responsible for first commercialization. The table indicates that, over time, the number of innovations per year increased. On average, between the years 1824 to 1926, we note an innovation coming along every seven years or so. In contrast, about one notable new material technology emerged annually over the following period beginning in 1934. Moreover, the United States increasingly dominated innovation. Whereas during that first period, 65% of new technologies originated from within the United States, a full 80% were American made during the period 1934–1964.

This pace of American innovation did not last, however. Beginning in the 1960s, a slowdown took place, a relative period of quiet after years of frenetic activity. It would be a good 15 to 20 years before the world would see its next—the third—technological revolution in materials, but when it came it opened up a universe of innovation possibilities that promised to dwarf the previous achievements of the

TABLE 1.1. The first two waves of materials: 1808–1964

1824	Portland cement invented (USA
1839	Vulcanization of rubber (United States)
1860	Stainless steel (United States)
1863	Synthetic dyes (United Kingdom)
1865	Celluloid (artificial plastic) (United States)
1870s	Coal-tar synthetics (Germany)
1886	Aluminum (Charles Hall, United States)
1906	Age hardening of aluminum alloy (United Kingdom)
1907	Bakelite (first entirely synthetic plastic) (United States)
1910	X-Ray crystallography (William Bragg and Max von Laue, United Kingdom)
1912	Synthetic ammonia—Haber process (Germany)
1913	Stainless steel "rediscovered" (United Kingdom)
1915	Pyrex™ (Corning, USA)
1923	Synthetic ammonia, "American" process (United States)
1925	Leaded gasoline (United States)
1925	18/8 austenitic grade steel adopted by chemical industry (United States)
1926	The first ethylene-based synthetics (United States)
1934	Nylon invented (United States)
1935	Styrene (United States)
1936	Clear, strong plastic (Plexiglass™) (United States)
1938	Fixed-bed catalytic cracking (United States)
1939	Polyethylenes (United Kingdom)
1941	Polyesters (United Kingdom)
1942	Fluid catalytic cracking (United States)
1943	Synthetic rubber (Collaboration) (United States)
1943	Saran™ (vinyl-based) (United States)
1944	Aviation gasoline (United States)
1945	Barium titanate ceramics (United States)
1946	Polyesters (United Kingdom)
1946	Poly T (Tupperware™) (DuPont, USA).
1947	Nickel-based superalloys (United States)
1947	Transistor (United States)
1949	Synthetic BTX (United States)
1949	Ceramic magnets (Netherlands)
1950	Synthetic glycerin (United States)
1953	Polycarbonate plastics (United States)
1953	High-density polyetheylene (Germany)
1953	Dacron™ (DuPont, USA)
1954	Synthetic diamonds (United States)
1954	Synthetic zeolites (United States)
1955	Teflon™ (DuPont, USA)
1955	High-molecular-weight propylene (Italy)
1957	Glass into fine-grained ceramics (United States)
1958	Lycra spandex™ (United States)
1960	Polyurethane (United States)
1960s	Silcone plastics (United States)
1964	Acrylic paints (United States)

Source: [2].

past. Paradoxically, this gale of creative destruction evolved from the existing and ever-nurturing technological landscape.

CONTINUITY AND NEW DIRECTIONS: 1980s AND 1990s

Beginning in the mid-1950s, American technology entered a period of stagnation, certainly relative to the dynamic period from 1925 through 1955. Incremental change rather than pioneering innovation characterized the period. This slack time for innovation lasted approximately twenty years. But by the mid-1970s, American technology saw the initial stirrings of a new-materials revolution, the third phase of modern materials development. This technological push has been accelerating its pace ever since. Just as the second materials revolution evolved from but moved far beyond the first, so this third revolution intersects with, but also radically diverges from, earlier movements in a number of ways. Certainly, we discern important technology transfer from established industry, such as high-pressure catalytic techniques and the use of existing materials such as advanced engineering plastics, thin organic films, and biotechnology, as the basis for the new-generation materials. This transfer of information and technology from an earlier revolution to the next can be thought of, as Martha Trescott believes, as "people transfer" [7]. Thus, we find that personnel, and often chemical engineers, move physically by various routes from the traditional chemical industry to departments and companies working in more cutting edge areas of new materials, bringing their experiences and knowledge with them. These agents then act as bridges linking the previous materials revolution with today's advanced-materials creation.

Although the evolutionary model of technical change is useful to keep in mind, especially if one is searching for the roots of the new-materials industry, it is a mistake to think that this pioneering effort is merely an "add-on" to, or incremental continuation of, the earlier achievements in plastics, rubber, and fuel. The break with the past is as, if not more, important to understand if we are to understand this new world of materials and its current and future impact on society. Indeed, to make a basic distinction between the past and present it is essential to define and establish the boundaries of what these new materials, so apparently diverse, have in common. From the 1930s through the 1960s, the universe of new materials operated on a common theme: the building up of so-called "macrostructures" by linking together molecular units, found in refinery off-gases, into super-long chains possessing desired physical and chemical properties.

The new order of technology creation rejects this once dominant axiom of advanced-material development. The older technology of the superpolymer clearly had been taken as far as it could, and impressively so. But if new worlds are to be conquered, and markets extended, new paths need to be found. Since the late 1970s, this pioneering route led to the creation, manipulation, and reconfiguration of very small molecular, and even atomic, units within a wide range of material categories.

In essence, this new way depends on the customization of atomic structure. These very small units, or micro building blocks, are often in the nano-sized range, but not necessarily so. A number of new materials that have been, or are currently being, developed for commercial application consist of larger than nano-scaled units but are still far smaller than the macromolecular chains of the past. From this, we see that the advanced materials of today are composed of fundamental units that are nano-scaled or larger but no more that a few linked molecules in length. Nanotechnology, then, contains, but is not the total universe of, this new generation of materials. The Table 1.2 displays the major achievements in these new materials through 2002. It is clear that the pace of innovation accelerated after the 1970s.

What are the implications of this realignment of focus from very large to far smaller units, especially in understanding the difference between past and present materials innovation? The greater flexibility and configurational and structural possibilities inherent in handling smaller building blocks is the essential characteristic that distinguishes this most recent technology. On a very fundamental level, the sheer number and variety of new materials currently in play or on the horizon is unprecedented, even compared to the technically active post World War II decades. Whereas the annual revenue from new materials commercialized in the 1950s globally was tens of millions (current dollars) in a given year, industry experts do not flinch at the prospect that sales of new materials and their products might very well hit the $1 trillion mark in only a few years, when the new materials revolution will still be relatively young [4]. Whether this figure is realistic is unclear, but the fact that it appears reachable to seasoned experts in industry and government speaks to the believed potential of a highly supple industry capable of producing a steady stream of products of a volume and quality not matched by either the German coal-tar machine of the late nineteenth century or the U.S. petrochemical juggernaut of the mid-twentieth century.

It is instructive to highlight the fundamental distinctions between the three periods or "waves" of advanced materials technology (see Table 1.3). Over time, we note an increasing flexibility and range of innovation. The first phase of innovation (1850 to 1930) engaged a set of technologies with only relatively limited room for expansion before diseconomies of specialization kicked in. There are only so many inorganic reactions possible, due to the nature of the internal structures involved. These noncatalytic reactions did create the very important metals revolution that so critically transformed the American economy after the Civil War, but the number of product innovations could not extend much beyond these relatively simple materials.

The second materials revolution, though clearly evolving from the earlier period, was different in many ways. As any chemistry student can testify, the number and complexity of organic reactions greatly exceed those possible within the inorganic (non-carbon-based) universe. The synthesis, or joining together, of simpler molecular units into long-chained leviathans was the key to economic possibilities. The linking of catalysts and petroleum feedstock to organic technology in the 1930s proved very felicitous to America's economic fortunes. This combination expanded mightily the number and range of possible syntheses and, therefore, man-made products capable of diffusing into and fundamentally augmenting society's most valued industrial activities. But, at the same time, such profusion of economically

TABLE 1.2. Landmarks in new materials

1950s	Germanium-based semiconductors
1950s	Diffusion furnace to diffuse dopants into silicon wafers
1959	First mention of possibility of fabricating materials atom by atom
1960	Large single crystals of silicon growth
1960	Magnetic cards (for computers)
1960s	Directionally solidified (DS) "super alloys"
1962	Nickel–titanium (Ni–Ti) alloy shape memory
1962	Semiconductor materials: slicing and doping of silicon crystals
1964	Carbon fiber
1964	Semiconductor material for circuits
1965	Thin-film resistor materials
1965	Multilayer metallization
1960s	Advanced composites: high-modulus whiskers and filaments
1970	Optical fibers
1970s	Single-crystal (SC) "super alloys"
1970s	Microalloyed steel
1970s	Amorphous metal alloys
1973	Kevlar™ plastic
1974	Metal matrix composites
1974	First molecular electronic device patent
1975	Solid-source molecular beam epitaxy (MBE)
1970s	Metal and polymer–metal composites
	Boron filaments
	Silicon carbide fibers
	Graphite-reinforced composites
1977	Electrically conducting organic polymers.
1980	New generation of deep UV-photoresist materials (for advanced computer lithographic techniques using "chemical amplification techniques")
1980s	Rare earth metals
1980s	Lanthanum–barium–copper oxide materials discovered to be superconducting
1982	Scanning tunneling microscope—atomic- and molecular-scale imaging (United States)
1984	Advanced polymer materials used for encapsulation of drugs for optimal delivery of medicines
1985	Gas-source molecular beam epitaxy (MBE) for thin films
1985	"Buckeyball" Fullerenes discovered
1986	New generation of optical polymers for flat-panel, liquid-crystal displays
1987	New generation of piezoelectric crystals
1987	Advanced electromagnetic materials for new generation of MRI technology
1988	New generation of advanced stainless steels for infrastructure
	Power generation
1990	Synthetic skin
1989	Tip of scanning tunneling microscope precisely positions 35 xenon atoms to spell "IBM"
1991	Carbon nanotubes discovered
1996	Vacuum arc–vacuum reduction stainless steel technology
1996	High-purity single-walled nanotubes (via laser vaporization)

TABLE 1.2. *Continued*

2000	Thin-strip casting (stainless steel)
2001	Superconducting "Buckminsterfullerene" crystals
2002	New generation of ultra-thin-layer, high-dielectric insulating materials
2004	Rise of the first commercial biorefineries in the United States

Source: [3].

useful innovation could not proceed indefinitely. Even with the use of catalysts—and only a certain number of commercially viable catalysts existed in any case—inherent rigidities in the nature of petrochemical synthesis seriously threatened a continual stream of technological growth. Simply put, there were only so many macromolecular products that could be made from petroleum sources. Although far more malleable and "giving" than the less technically elastic coal (as the Germans would find out by the 1920s to their dismay, as they lost their once considerable world chemical might to the United States), petroleum and natural gas were just as susceptible to eventual diseconomies of specialization and decreasing returns to scale that this implies. By the mid-1950s, the R&D costs to extract new building blocks from petroleum and natural gas and link them together into ever larger molecular chains with fundamentally new and economically potent properties increased exponentially. At the same time, petrochemical companies, now more closely controlled by financial players with their attention to stock prices and quick-profit leveraged buyout opportunities, displayed an unprecedented aversion to the growing risks that have so famously characterized modern chemical corporate R&D [5].

By the late 1970s, it was clear to many in the chemical industry that the next

TABLE 1.3. Characteristics of the major advanced materials revolutions: 1850–present

	Field	Scale	Process	Production	Raw Materials	Technological impact on society (% of economic growth accounted for by the new materials)
First materials revolution: (1850–1930)	Inorganic	Atomic	Oxidation/ reduction	Thermal	Coal; ore	25%–45%
Second materials revolution: (1930–1960)	Organic	Macromolecular	Synthesis	Catalytic	Petroleum; natural gas	50%–65%
Third materials revolution: (1980–present) other	Inorganic and organic	Nanoscale to a few molecules in length	Reconfig-urational	Semicatalytic	Coal; petroleum; natural gas; agrimaterials;	65%–90%

Source: [2, 3].

wave of innovation would have to come by finding a middle ground between the characteristics intrinsic to the first and second material revolutions. Expanding technological flexibility could no longer come from just inorganic or just carbon-based synthetics but had to incorporate both. This, in turn, could only happen by sidestepping the structural limitations and rigidities imbedded in macromolecular modeling. The structural units making up the newest materials had to be on a much smaller scale but cover a range of dimensional size from the nano-level up to units a few, and only a few, molecules in length. Rather than create new macromolecules out of the simpler molecules in fossil fuels, researchers began finding ways to design new and useful materials by rearranging the internal structures of individual molecules.

The breaking down of the "super" molecule into smaller parts vastly increases the structural possibilities of both inorganic and organic materials. It also opens up raw material possibilities and a wider range of reaction types and gives free reign to all sorts of inorganic–organic hybridization, thus adding a third dimension to materials innovation not possible in earlier periods. All of this stretches considerably the range of technological flexibility, the flow and type of possible products, and, ultimately, market development and growth.

THE NEW MATERIALS AND THE RISE OF THE "TECHNOLOGICAL" SOCIETY

When this new approach to technology creation first emerged in the 1980s, it could not have come to light at a more propitious time. Beginning in the nineteenth century, technology began to play an increasingly central role in economic performance. The linkage between technology and economic growth grew stronger in the twentieth century. Moreover, new-material technology progressively dominated the performance and even the very existence of a nation's technology overall and thus, in turn, of the rate and direction of its economic activity. Today, a new generation of materials plays a far greater role in determining industrial competitiveness than their counterparts did in the past. Certainly, steel, aluminum, and the first synthetic materials diffused into and helped advance the critical industries, including railroads, automobiles, aircraft, telecommunications, and defense. But these materials did not by themselves set the pace of innovation in these industries. The industrial revolution in the United States was a mechanical affair and technology depended first and foremost on advances in engineering design as well as materials. For example, mechanical expertise, electrical hardware, and the assembly line figured more prominently in the development of, respectively, the steam engine, telegraph, and automobile than did adoption of advanced materials. Even after World War II, the new synthetics merely substituted for older materials in the textiles, automotive, aerospace, and electric power industries. These incursions on the part of the new materials provided important benefits to the industries and society as a whole but did not cause, on their own, fundamental changes in how the technologies that incorporated these materials worked or in what directions they evolved.

The technical landscape shifted dramatically by the 1980s. Industrial competitiveness within a country, region, or area now depends on the continued development and growth of the so-called "science"-based industries, including biotechnology, pharmaceuticals, microelectronics, chemical synthesis, and energy. But advances here, such as new IT technology, drug delivery systems, organ replacement, genetic engineering, organic electronics, solar cells, and so forth, rely directly and often exclusively on the development of new materials, in one form or another. Scientists, for example, expect the silicon chip to reach its technological limit in its ability to miniaturize. When this point is reached, further advance in miniaturization will depend exclusively on adapting new electronic materials. In this sector, for example, advances in electronics have depended on new photoresist substances and on future progress in organic conducting polymers. No longer the mere handmaiden of industry, the materials sector sets the pace and determines the direction of technological change in the most dynamic industries within the global economy [6].

But what exactly is the nature of this impact? If a technology is to influence the economy in any way, it must, as any product, do so through the mechanism of the marketplace. Now that we have learned something of the context for new materials, we will turn our attention to the market for these materials. In doing so, we concern ourselves with the recent past (from the 1970s), the current market profile, and future projected impacts up to 2030.

REFERENCES

1. Good, M. L. (2007), "High-Performance Materials," Little Rock, Arkansas: University of Arkansas (www.greatacheivements.org), p. 1.
2. Time line constructed from a variety of sources, including, Spitz, P. H. (1989), *Petrochemicals: The Rise of an Industry,* New York: Wiley; Good, M. L. (2007), "High-Performance Materials," Little Rock, Arkansas: University of Arkansas (www.greatacheivements.org), p. 1; Plotkin, J. S. (2003), "Petrochemical Technology Developments," in Spitz, P. H. (Ed.), *The Chemical Industry at the Millenium: Maturity, Restructuring, and Globalization,* Philadelphia: Chemical Heritage Press, pp. 51–84; Aftalion, F. (2001), *A History of the International Chemical Industry: From the "Early Days" to 2000,* Philadelphia: Chemical Heritage Press.
3. Timeline is reconstructed from a variety of sources, including Good, M. L. (2007), "High-Performance Materials," Little Rock, Arkansas: University of Arkansas (www.greatacheivements.org), p. 1; National Science Foundation (NSF) (2005), *Advanced Materials: The Stuff Dreams are Made of,* National Science Foundation; Washington, D.C., The Evident Technologies Website (www.evidenttech.com), Thayer, A. M. (2003), "Nanomaterials," *Chemical & Engineering News,* Vol. 81, No. 35, pp. 15–22; Holister, P. (2002), *Nanotech: The Tiny Revolution,* Las Rozas, Spain: CMP Cientifica; *Nanoinvestor News* (www.nanoinvestornews.com), 2001–2005; Moskowitz, S. L. (2002), *Critical Advanced Materials Report: Final Draft,* Division of Advanced Materials and Electronics, Virginia's Center for Innovative Technology (CIT), Herndon, VA; Teresko, J. (2003), "The Next Material World," Industryweek.com, April 1.
4. For a discussion of the various forecasts made for the global nanotechnology market and the issue of definition of "nanotechnology," see European Nanoforum (2007), *Nanotech-*

nology in Europe—Ensuring the EU Competes Effectively on the World Stage, Survey & Workshop (Dusseldorf, Germany), June 21, p. 9.

5. Da Rin, M. (1998), "Finance and the Chemical Industry," in Arora, A., Landau, R., and Rosenberg, N. (Eds.), *Chemicals and Long-Term Economic Growth: Insights from the Chemical Industry,* New York: Wiley, pp. 307–339; Roberts, J. (2003), "The Financial Community Takes Charge," in Spitz, P. H. (Ed.), *The Chemical Industry at the Millennium,* pp. 283–310.
6. For a discussion of the growing role of chemicals and advanced materials in electronics and semiconductors, see Brock, D. C., "Reflections on Moore's Law" (2006), in Brock, D. C. (Ed.), *Understanding Moore's Law: Four Decades of Innovation,* Philadelphia: Chemical Heritage Press, pp. 87–108.
7. Trescott, M. (1981), *The Rise of the American Electrochemicals Industry, 1880–1910,* Westport, Connecticut: Greenwood Press.

Part Two

Opportunities and Risks

Chapter 2

A Great Potential—Markets and Society

What exactly is it about this group of advanced materials that makes them so important to society? In economic terms, what has been the market impact of these materials within the United States and internationally and what can we expect their impact to be as these products and processes continue to be developed, refined, and commercialized? The previous chapter introduced us to these materials and provided a useful historical context. This is fine, as far as it goes. But those discussions gave us little feel for the materials themselves, what they are exactly and why they present such promise to so many industries and the global economy as a whole. This chapter aims to peer into the details of these materials a little more deeply; to, at the least, get a sense of the economic potential of these technologies, especially their broad usage and relevance to modern society. The analysis considered here assumes an ideal world of convergence in which globalization forces a progressive coming together of economic activity among nations. Thus, we hypothesize a "best of all possible worlds" scenario—a hypothesis that we will need to test later on—that, over time, Europe will attain the very same advantage in creating new technology and capturing markets for these technologies that the United States enjoys.

A beginning point to our further understanding of the products themselves is to note that, at present, certain of these materials have achieved dominant markets, most notably, nanoceramics and advanced metals (such as nanometals and superalloys). For example, the material nano-silica is used in so-called "planerization" slurries to smooth chips for the semiconductor industry. The more advanced stainless steels have, of course, found their way into a host of industries (e.g., power generation, chemicals, and so forth), and nanotitanium dioxide is a familiar component in the cosmetics industry. New nanomaterials have also led to advances in computer disc drive and data storage, solar cells, and rechargeable batteries. But our interest extends beyond these materials to those products and processes that are just emerging as commercial products and that are likely to grow in importance over the next decade and a half and supply the critical foundation and springboard for the most important technologies of the twenty-first century. In others words, in this story we consider new materials of the near future as well as those of the recent past and those that are currently commercially active.

The following sections examine these essential advanced material areas. Industry specialists consider the products to be highlighted over the next few pages as the most important advanced materials that have recently entered, or will shortly be introduced to, the market, whether in the United States or internationally. As a group, these materials account for the bulk of current and projected advanced materials sales within the United States and worldwide for the 2005–2020 period.

The market arena relevant to advanced materials, as defined at any particular time, does not remain constant. Rather, a dynamic and evolutionary relationship exists between advanced materials and their market environments. As advanced materials science and technology progresses, and as the ability of industry to manipulate and even design internal molecular structures proceeds, the range of application expands into previously untouched commercial venues. The following tables illustrate this by summarizing the major markets for each of the important advanced material groups.

The clearly greater number of "boxes" marked off in the lower part of Table 2.1 shows the greater range of applications of the newest generation of advanced materials. A rudimentary accounting of the matrix confirms this impression. The most important materials that came to market during the first six decades of the twentieth century can claim a total of 24 applications, or approximately 2.7 applications per product. How much more dynamic the post-1980 period seems in comparison. The 14 major new products listed here account for 120 applications in total, or an average 8.6 applications per material. In other words, since the 1980s, the most recent advanced materials have been diffusing into the economy, if we measure diffusion by applications per material innovation, at approximately three times the rate as did the new materials of the previous generation. Given this reality of market penetration by modern material technology, robust estimates of worldwide market sales in the hundreds of billions of dollars for advanced materials as a group may in fact be a reasonable prediction, especially as these materials become ever more the sine qua non of progress and growth in the basket of industries that define a modern economy. Table 2.2 lists the major industries that consume advanced materials.

But what exactly are these new materials? What is it about them that makes them actually and potentially such a potent market force in modern society? We have discussed in the most general terms their structural characteristics, but it is pertinent to delve in greater detail into the particularities of the major product groups and to focus on their more important market applications. These discussions are far from complete as they do not include all the advanced materials currently in progress toward market, nor do they encompass the complete pantheon of applications for particular product categories. Nevertheless, they highlight the essential contours of what is today the advanced materials landscape.

TABLE 2.1 The market matrix for advanced materials

Time period	Company	Technology	Chemical production	Construction	Textiles/ apparel	Auto-motive	Infra-structure	Aero-space	Electrical equipment	Bio-technology	Agriculture	Electronics	Energy	Military/ defense	Environ-mental
I. 1900–1930s	Union Carbide	Electrometals		X		X	X	X	X				X	X	
	Alcoa	Aluminum	X	X		X	X	X					X	X	
	DuPont	Synthetic ammonia (nitrates)	X								X				
II. 1940s–1980s	Union Carbide	Polyethelenes		X		X									
	DuPont	Nylon		X	X										
	Dow Chemical	Styrene, polystyrene													
	Sun Oil	Fuels				X									
	Exxon	Fuels, synthetic rubber				X			X						
	ICI	Polyesters			X										
III. 1980s–Present	Applied Nanoworks	Nanocrystals	X			X	X			X	X	X	X	X	X
	Arrow Research Corp.	Nanocomposites		X		X	X	X	X	X		X	X	X	X
	Carbon Nanotechnologies	Nanotubes	X			X	X	X	X	X		X	X	X	
	Cargill Dow	Biosynthetics	X	X	X	X	X	X	X	X	X	X	X		
	Evident . Technologies, Inc	Nanocrystals/ quantum dots	X							X		X	X		
	InMat Inc.	Nanocomposite coatings		X		X	X	X	X	X		X	X	X	
	Lucent Technologies	Organic electronic materials		X	X	X	X	X			X	X	X	X	
	Lumera Corp.	Advanced polymers, electro-optical polymers	X	X	X	X	X		X			X		X	X

(*continued*)

TABLE 2.1 *Continued*

Time period	Company	Technology	Chemical production	Construction	Textiles/ apparel	Auto-motive	Infra-structure	Aero-space	Electrical equipment	Bio-technology	Agriculture	Electronics	Energy	Military/ defense	Environ-mental
III. 1980s–Present (*cont.*)	Nano-Tex	Advanced fibers		X	X		X	X	X			X	X	X	X
	Nanodynamics, Inc.	Advanced ceramics, nanotubes	X	X		X	X	X	X	X	X	X	X	X	X
	Nanofilm, Ltd.	Ultrathin coatings	X	X	X	X	X	X	X	X		X	X	X	
	NanoProducts Corp.	Catalysts/ nanopowders													
	Nanotechnologies, Inc.	Nanopowders	X	X	X	X	X	X	X	X		X	X	X	X
	Polycore Corp.	Nanocomposites, nanoclay	X	X	X	X	X	X	X	X		X	X	X	X

Source: [1].

TABLE 2.2. The advanced materials consuming sectors

Biotechnology
Pharmaceuticals
Computers and telecommunications (nonconsumer)
Automotive
Aerospace
Construction and infrastructure
Private security
Industrial equipment
Industrial intermediates
Packaging
Electronic consumer products
Nonelectronic consumer products
National security
Service industries: high technology
Textiles and apparel

THE ADVANCED-MATERIAL FAMILIES: CHARACTERISTICS, TECHNOLOGY, AND APPLICATIONS

Bioengineered Materials [2]

Biochemicals play an increasingly critical role in the advanced materials industry. An important biochemical technology that is just beginning commercialization is the so-called bioengineered materials. These materials bridge the biochemical and synthetic organic fields and are expected to provide significant volumes of synthetic materials over the next few years.

Biorefinery technology involves biochemical transformation, in so-called "biorefineries," of agricultural feedstock, by-products, and wastes into useful synthetic materials. These products include synthetic plastics and packaging, clothing, fuel additives, chemicals (e.g., alcohols, polymers, ethylene, phenolics, acetic acid), biologics, food products, adhesives and sealants, and commodity and industrial products. As with most biochemical production, these processes are generally batch or semicontinuous in nature. Attempts are being made to reach closer to fully continuous operations, which will lower manufacturing costs. As markets expand, the scaling-up process must accelerate, which, in turn, creates economies of scale through application of chemical engineering techniques and automation systems. These include developments in large fluidized-bed bioreactor systems and modular setups that speed the scale-up of processes. New design of bioprocess separations technology (including microfiltration) continues to advance the state of the art in biorefining production.

Research and development in this area is proceeding worldwide by multinational chemical and material-processing companies. Within the United States, established companies such as Dow Chemical (through a joint partnership with Cargill Corp.)

pursue research and development in this field. Dow–Cargill has itself commercialized one of the first such plants in the country. In particular, the company develops processes, based on corn stock, to provide the building blocks for a variety of synthetic products, including fiber and plastic materials with superior characteristics, such as wear resistance and insulation, useful in clothing, furnishings, containers, packaging, and numerous industrial applications.

It is expected that, with the financial backing of some of the large chemical companies and the growing innovation of smaller and medium-sized companies that are coming into the field, bioengineered-materials producers will find markets for their products over the next ten years. Greater volumes, lower prices, and higher quality products will mean greater competitive positioning against traditional petrosynthetics. By 2012, it is estimated that total U.S. sales of these materials could reach in excess of $1 billion. On a worldwide basis, sales could exceed $3 billion by this time.

Advanced Metals: Advanced Stainless Steel and "Superalloys" [3]

Both stainless steel and the industrial nonferrous metals (e.g., aluminum) first appeared as commercial products in the late nineteenth century. Incremental improvements in alloy quality and in production processes proceeded over the next half century. By the 1950s and 1960s, a new group of advanced metal products collectively called "superalloys" entered the marketplace.

The term was actually first used shortly after WW II to describe a group of alloys developed for use in aircraft turbine engines that required stainless steels that could perform well at very high temperatures. Since then, a number of superalloys have emerged commercially. These alloys, containing chromium, cobalt, nickel, manganese, molybdenum, and vanadium, came online during the 1960s and 1970s. These metals demonstrate a range of superior properties, depending on chemical constitution and microstructural configuration. For example, they withstand fierce heat while spinning in jet engine turbines. Some of these metals possess exceptional strength and hardness, others are virtually impervious to corrosion or withstand constant flexing or have desirable electrical or magnetic properties.

By the 1980s, advances in the superalloys extended the range of market possibilities. New processing technology led to the making of advanced metal parts grown from single crystals in ceramic molds. This technique allowed more intricate parts to be manufactured to more precise specifications. By this time as well, a new type of superalloy entered the scene—the so-called "6 Mo superaustenitics." These stainless steel alloys containing 6% molybdenum excel in corrosion resistance and are produced at a much lower cost than the traditional (pre-1980) nickel-based alloys.

Superalloy production requires new types of process technology. An important advance is the vacuum induction melt (VIM) furnace that provides stringent metallurgical cleanliness and exacting chemical control. In this technology, raw material is melted in an airtight vessel with the pressure carefully regulated. Since the metal does not come into contact with atmospheric contamination, the process allows better control over alloy chemistry. Both the United States and Germany pioneered this technology in the 1980s and 1990s. The United States excelled in scaling up the process for efficient, low-cost operations. One of the first U.S. companies to apply the process on a commercial scale, Carpenter Technology Corp., in a typically American fashion, radically modified a German-made VIM process to make advanced alloys more efficiently.

Further downstream, advanced processing techniques for superalloys advanced significantly over the last half century, from directionally solidified (DS) techniques in the 1960s to single-crystal (SC) technology in the 1970s and 1980s. In DS processing, columnar grains form parallel to the growth axis. SC casting is similar to the DS technique but a single grain is selected. The exceptional properties of both DS and SC alloys result from the alignment or elimination of any weak grain boundaries oriented transverse to the eventual loading direction. However, in contrast to DS technology, the SC method uses a solidification process by which a single grain grows to encompass the entire part. This results in outstanding strength and high melting point through the elimination of grain boundaries that are present in DS materials. Unlike the DS route, the homogenization heat-treatment temperature is increased without fear of incipient melting, thus allowing for more complete "solutioning," an important criteria for high-quality alloys.

Advanced Ceramics and Superconductors [4]

As with the superalloys, the advanced-ceramics group illustrates how present-day molecular manipulation promises wide-scale application within the inorganic field. This is of enormous significance in that it reintroduces the inorganics as a cutting edge domain of activity, which has not been the case since the 1920s.

Ceramics are, in general, nonmetallic, crystalline materials. There are three major forms of ceramics: amorphous, polycrystalline, and single crystals. Ceramics are generally made from powders and additives under high temperatures. The traditional ceramics are very common materials in the construction and crafts industries. These include bricks, tile, enamels, refractories, glassware, and porcelain.

It is the advanced-ceramic field that leads us into entirely new technologies and expanded market opportunities. An "advanced" ceramic is an inorganic, nonmetallic, basically crystalline material of carefully controlled composition and manufactured from highly refined raw materials giving precisely specialized attributes. In contrast to the traditional ceramics, advanced ceramics are fundamentally crystalline substances with highly engineered microstructures in which grain size and shape, porosity, and phase distribution are carefully planned and controlled. Internally, they possess very small grain sizes to achieve high surface area contacts. These materials require detailed regulation of composition and production, such as the use of pure, high-quality synthetic compounds as inputs and a clean-room environment for processing. These materials exhibit unique or superior functional attributes, such as great strength or conductivity.

Advanced-ceramics technology subsumes the relatively new field of metal powders. Like traditional ceramics, advanced ceramics are densified from powders by applying heat, a process known as sintering. Unlike traditional ceramics, advanced powders are not bonded by the particle-dissolving action of glassy liquids that appears at high temperatures. Instead, solid-state sintering predominates. In this process, matter from adjacent particles, under heat and pressure, diffuses to regions that grow between the particles and ultimately bond the particles together. As the boundaries between grains grow, porosity progressively decreases until, in a final stage, pores close off and are no longer interconnected. Since no glassy phase is needed in solid-state sintering to bond particles, there is no residual glass at the grain boundaries of the resulting dense ceramic that would degrade its properties. The use of specially processed dopants that are applied to the general metal powder's internal structures allows a greater range of materials to be made. Novel high temperature and pressure processes force the metal interlopers into a strong binding arrangement at the atomic and molecular levels that impart superior physical, mechanical, and electrical properties.

Accordingly, advanced ceramics can have a broader variety of applications that were in the past well out of the range of the traditional ceramic materials. One of these major markets is in electronics, which accounts for approximately 66% of the total demand for advanced ceramics. Important ceramic products in electronics applications include both the pure and mixed oxides—alumina, zirconia, silica, and ferrites—and doped barium and lead titanates. The important electronics application of these materials touch the very heart of Moore's Law and the ability of a country's electronics industries to maintain their rigid requirements of advancing capacities, smaller dimensions, and lower costs. The advanced ceramics find their particular strengths in substrates and packaging, capacitors, transformers, inductors, and piezoelectric devices and sensors. Approximately two-thirds of total electronic ceramic consumption goes into integrated-circuit packages and capacitors.

Advanced ceramics find important application in the construction, automotive, machine tool, and industrial equipment industries due to their resistance to corrosion and high temperatures and ability to withstand mechanical stresses and strain. Ceramic–metal matrix composites that incorporate reinforcing materials such as carbon fibers are particularly important in the construction and related industries. These materials possess superior mechanical properties, excellent thermal stability, and a low friction coefficient (allowing them to serve as superior lubricants). Examples of such materials include silicon carbide fibers in silicon carbide matrices and aluminum oxide fibers in aluminum oxide matrices. Important advanced structural ceramics include various forms of aluminum oxide, zirconia, silicon carbide, and silicon nitride. Such ceramics are important materials for infrastructure applications, such as power plants, construction, and bridges, as well as in industrial equipment, for example, bearings, seals, and cutting tools. The automotive industry employs advanced structural ceramics in catalytic converters and for certain under-the-hood components including spark plug insulators, catalysts and catalyst supports for emission control devices, sensors of various kinds, and engine parts.

A subset of these types of ceramic materials is advanced ceramic coatings and powders. Ceramic-coatings technology deposits a thin layer of ceramic on complex surfaces at low cost for improved resistance to corrosion, mechanical wear, and thermal shock. Currently, gas turbine engines use zirconium oxide coating technology, which provides a thermal barrier that allows engines to run hotter by protecting the underlying metal. In turn, the coating extends component life, increases engine efficiency, and reduces fuel consumption. Research in advanced-ceramic coatings focuses on improving adhesion of the coatings to surfaces, improving the properties of the coatings, and reducing the costs of the coating process. Recent developments in coating include physical- and chemical-vapor deposition (PVD and CVD). In PVD, for example, a high-energy laser blasts material from a target and through a vapor to a substrate, where the ceramic material is deposited. Even more precise control over the deposition of thin ceramic films requires the use of molecular-beam epitaxy (MBE), by which molecular beams are directed at and react with other molecular beams at the substrate surface to produce atomic layer-by-layer deposition of the ceramic. Ceramic powder processing is a major field in itself. Particular applications either currently used or close to realization include very fine abrasives for chemical mechanical polishing, advanced catalysts for solid rocket fuels, magnetic recording media, optical fiber coatings, magnetic materials as well as application in a new generation of fuel cells, oxygen sensors, optoelectronic devices, and ceramic structural components.

Research and development worldwide has also been accelerating in the field of ceramic superconductors. These materials have the ability to carry electric current with zero resistance and so do not have to release any energy as heat. In the 1980s, IBM advanced the field in their work on lanthanum–barium–copper oxide superconducting ceramics. It has also been found that ceramics based upon mixtures of indium oxide and indium tin oxide (ITO) as well as semimetallic ceramics (e.g., the oxides of lead and ruthenium, bismuth ruthenate, and bismuth iridate) make excellent conductors that are also optically transparent. Superconducting ceramics provide a number of advantages, including low power dissipation, high operating speed, and extreme sensitivity. Applications for superconducting ceramics include magnetic imaging devices; energy storage; solar cells for liquid crystal displays (LCDs), such as those used in laptops; and thin-film resistors in integrated circuits.

The market potential of the advanced-ceramics group cannot be more clearly expressed than in the growth of two major advanced-ceramic material fields: nanoceramics and piezoelectric ceramics. These materials are the focus of much research and development activity both within the United States and globally. They are, in fact, currently poised to begin making major incursions into their major markets. These materials demark the cutting edge in the advanced-ceramics field.

Nanoceramics

This technology is distinguished by the fact that its nanoceramics-based metallic units are far smaller than is the case with other ceramic-powder systems. Such materials, processed under intense heat and elevated pressure, have the singular ability to form parts and components that extend performance capabilities well beyond current limits. The properties and applications of nanoceramic materials generally depend on the type and average particle size range of these metal structures. The first nanceramic materials to see market activity include titanium nitride, silicon nitride, aluminum nitride, zirconia (and zirconia–aluminua), yittrium–aluminum compounds, and ceria and gallium oxides.

Nanoceramics lead to significant cost savings and new material applications. Traditional ceramics tend to be hard and brittle, thus making parts made of the material difficult to machine, and, in turn, significantly limiting markets. Nanoceramics, characterized by very small internal grain size, impart to products a mechanical flexibility that allows greater ease in forming, shaping, and finishing (e.g., grinding and polishing) in lower temperature environments. Nanoceramics also possess superior structural characteristics exhibiting high strength and excellent abrasion, deformation, and wear resistance, even at high temperatures. Cost savings to industry result from lower energy use, reduced time to complete operations, and material savings from fewer damaged parts requiring replacement.

Applications for nanoceramics include structural and industrial uses, such as in machine tools, electroplated hard coatings, and thermal barrier coatings. In the automotive area, nanoceramics are used in "under-the-hood" applications (e.g., automotive engine cylinders providing greater retention of heat and more complete and efficient combustion of fuel). Small, light-weight sensors made of nanoceramics help to measure air/fuel ratios in exhaust gases. This in turn leads to more efficient cars and aids in curtailing environmental pollutants. Nanoceramics continue to diffuse into such industries as appliances, industrial machinery, and petrochemical and power plants. Nanoceramics should see growing application as well in liners and components for appliances, heat exchange systems, industrial sensors, electric motor shafts, gears and spindles, high-strength springs, ball bearings, and, potentially, thousands of additional structural parts and components.

But nanoceramics are not limited to the mechanical realm. Due to their unique internal structure—they possess a large number of molecular-sized "holes"—they serve as advanced molecular sieves and catalytic carriers for chemical, refining, and biotechnology operations. Because of their generally nonhazardous nature, chemical inactivity, and biocompatibility, nanoceramics supply materials for industrial, chemical, and biochemical ultrafiltration equipment, "delivery" systems that more efficiently and effectively introduce bioactive agents into the body, and equipment and apparatus for chemical and biochemical research and manufacture. Promising biotechnology applications include the use of nanoceramics in new bone-implantation systems and as implanted medical prostheses.

Nanoceramic materials also possess superior electro-optical properties that have applications as materials for semiconductors, electronic components, and related technology and systems, including optical filters; capacitors; floppy discs, tapes, and other magnetic media; fiber optics; and superconducting products (e.g., flexible superconducing wire). Nanoceramics, in the form of alumina, ceria, zirconia, and titania oxides, improve industrial micropolishing operations since their ultrafine abrasive particles provide superior mechanical polishing of dielectric and metallic layers deposited on silicon wafers. Nanoceramic materials demonstrate unique optical properties useful in advanced lighting systems. In particular, translucent alumina-based ceramic tubes can operate in high-pressure sodium lamps and metal halide lamp tubes for indoor lighting. Nanoceramics will also find increased use in critical energy-related technology, including advanced fuel and solar cells and new-generation microbatteries.

Table 2.3 shows the current and projected distribution of demand for nanoceramics in major markets. As is seen, currently, the largest share (53%) of the nanoceramic market goes into electro-optical applications, followed by chemical/environmental applications (40%). The remaining 7% of the market for nanoceramics entered into a variety of structural applications. Over the next few years, it is expected that the electro-optical and structural areas will gain ground at the expense of chemical and environmental applications.

Piezoelectric Ceramics

Piezoelectric materials—composed of mixtures or complexes of zirconium, titanium, lead, and other metals—create driving voltages when placed under mechanical stress (the "generator effect") and undergo mechanical movement or deformation when subjected to electrical impulses (the "motor effect"). There are four types of piezoelectric materials: ceramics, crystals (e.g., piezoelectric quartz), ceramic/polymer composites, and polymer films. Of these, piezoelectric ceramics represent the largest and most mature market segment, accounting for approximately 90% of the total piezoelectric market. If piezoceramic/polyer composites are included, then piezoelectric ceramic materials capture about 93% of the total piezoelectric market.

There are a number of piezoelectric ceramic materials, most composed of some form of lead or titanium. Currently, the most common piezoelectric ceramic is lead zirconate titanate (PZT). Other types of piezoelectric materials include barium titanate, bismuth titanate, lead titanate, and lead metaniobate.

TABLE 2.3. Distribution of demand for nanoceramics, by year

	2005	2010	2015	2020
Electronic/optical	53%	55%	56%	60%
Chemical/environmental	40%	37%	35%	30%
Structural	7%	8%	9%	10%

Source: [5].

Piezoelectric ceramics come in "bulk" and "multilayered" form. The bulk form of the ceramic consists of a single ceramic block from which are produced various shapes: blocks, plates, discs, cylinders, rods, and so on. In contrast, the multilayered variety consists of several thin layers of the ceramic material stacked up into rectangular and cylindrical shapes, such as bars, plates, and discs. The quality and performance level of these various forms of piezoelectric ceramics are measured by such variables as dielectric constants, dielectric loss factor, electromagnetic coupling factor, piezoelectric load constant, elastic compliance, elastic stiffness, electrical resistance, and thermal coefficient.

Piezoelectric ceramics are a growing presence within the biomedical, aerospace, automotive, industrial, consumer, and marine industries. While currently the government dominates the demand for U.S. produced piezoelectric ceramics, the industrial and consumer markets account for a growing proportion of piezoelectric ceramic sales. These materials enter into the making of such components as electric circuit elements, transformers, actuators, transducers, and energy generators (e.g., batteries). The components that are made of piezoelectric ceramics demonstrate a large force capability and short response time. This means that they provide rapid, precise, and carefully regulated displacement of devices, equipment, and systems in response to even small applied voltages. These piezoelectric ceramic components, in turn, find current and potential application in such devices and systems as sensors (medical, pressure, flow, and acceleration), sonar equipment and hydrophones, laser positioning, industrial tools and hardware (valves, meters, cutting and polishing machines, and displacement gauges), electrical devices (remote control switches, relay contact drivers, electro-acoustic devices, microposition actuators, electrical appliances, security alarms, and camera shutters), and security systems.

One area that is particularly promising for piezoelectric ceramics is their use in vibration control due to the general use of more powerful machinery and equipment in industry. Vibration control is especially important in such areas as aircraft, hospitals (e.g., vibrations due to MRI equipment), and power plants. Another potential market for piezoelectric ceramics is in the manufacture of wireless switching equipment for both residential and business structures. The market in Europe and Asia for these devices is growing due to the higher cost of installing and replacing wired systems in these regions. The specific uses for piezoelectric ceramics in nonwired applications include switching and lighting systems, appliances, security systems, doorbells, and burglar alarms.

Future applications of piezoelectric ceramics hinge on their superior power density and cost and size advantages. As a result, they become the material of choice for nonmagnetic transformer components in radio-frequency (RF) transmissions systems and remote control devices, power backlighting for computer screens and, in the form of ceramic fibers, as critical materials in the monitoring of stresses and strains in aircraft bodies, automotive engines, and building structures.

Investigation into more advanced piezoelectric materials and production processes continues apace within the United States and internationally. Attempts to find alternative materials that do not depend on lead result from stricter environmental policy. Denmark, for example, conducts research on the alkaline niobates as a possible substitute for PZT. Work is also underway in various countries on a new variety of porous piezoelectric ceramic that promotes superior performance for transducers operating underwater (e.g., in hydrophones). Finding new ways to fabricate piezoelectric ceramics is a priority. Significant work in the United States, England, and other countries centers on "net shape fabrication" and the process of "plasticizing" the powder–binder mixture in order to limit sintering and, in turn, the structural defects and high production costs associated with the sintering operation.

Synthetic Engineering (Nonconducting) Polymers [6]

Synthetic engineering polymers encompass the mature portion of the advanced-materials industry. Led by the great macromolecular discoveries of the post World War II period—nylon, polyester, polyethylene, the urethanes, the carbonates, and related polymers—the golden age for these materials climaxed two decades ago. Nevertheless, improvements and modifications continue to refresh these basic materials and even manage to breathe new economic life into them as they find extended usefulness in both new and familiar industries. Innovation here depends less on creating new and longer macromolecular polymers (this had already been done decades ago) than on manipulating and altering the smaller molecular units—the very building blocks—composing the structure. In this important sense, we can say that these new polymers are properly "advanced" materials, as we previously defined them. For example, polymer–metal composites (introduced in the 1980s) and, more recently, nanoenhanced polymers, are leading materials within the United States and internationally. These materials combine organic with non-carbon-based technology. As such, they find a wide range of applications including textiles, construction and infrastructure, electrical equipment, electronics, consumer goods, biomedical, and defense.

It is not appropriate, of course, to claim all synthetic polymers as advanced materials. The bulk of these are well known, or impacted, technology. It is estimated that out of every $100 of additional revenue generated by this sector between 2005 and 2015, approximately 18% can be attributed to cutting edge technology, for example, advanced composites and nano-enhanced polymers.

Advanced (as well as some traditional) polymers are already an important ingredient in electronics systems. Currently, in the United States alone, over one billion pounds of polymer materials are used in electronic applications, representing over $4 billion in sales. These nonconducting polymers find their markets in the so-called "passive" (nonelectric) applications, which include casings, adhesives, sealants, polarization layers, and substrates. These polymers are critical components in computer and information-based systems.

Organic electronic materials—those polymers that conduct electricity and can be fabricated as semiconductors—also come under the category of advanced synthetic engineering polymers. Because of their unique importance in the advanced-materials field, this group will now be discussed separately.

Organic Electronic Materials (Conducting Polymers) [7]

Organic conducting polymers (otherwise known as organic polymer electronic or OPE materials), are transforming many industries, with electronics and semiconductors serving as the root application. In this story of organic polymer electronics, the very existence of silicon as *the* material of the age can no longer be assumed as we peer into the not-so-distant future.

We are concerned here with the so-called "active" organics that perform actual electronic functions. There are two types of active electronic polymers: conductors and light emitters. With light emitters, a polymer generates light when electricity flows through it or it is exposed to an electric field. These organics perform the same function as, and can substitute for, phosphors, which are currently used in electroluminescent and organic light-emitting displays. A potential market for these light-emitting organic polymers is in light-emitting diodes (LEDs), whose active ingredient is a tiny diode made of a polymer rather than crystals of gallium arsenide or related material. The second type of active organics is termed polymer "conductors."*

Organic electronic materials possess the important property of relatively weak intermolecular bonds. This means that these materials behave both as semiconductor and electric insulators. A problem with OPEs however is their structural instability, especially in high-heat environments. Research undertaken by Eastman Kodak in the 1980s took a major step forward in addressing this difficulty by producing a multilayer organic device that was a close cousin of a p–n junction and that created efficient light emissions. Further work was undertaken in the 1990s by researchers in the United States, Japan, and the United Kingdom. Bell Labs' organic electronics research produced a number of breakthroughs in the 1990s, including n-type organic semiconductors, complementary circuits, and plastic matrix organic display backplanes, which are arrays of transistors that drive the pixels of displays and are considered important market opportunities by several firms within the United States and internationally. By the late 1990s, understanding the growing reality of OPE material technology, corporations such as 3M undertook comprehensive market studies in the field of organic electronic polymers and their applications. Table 2.4 shows the various OPE markets and materials that are either extant as of 2008 or being developed for introduction by 2012.

OPEs and Display Technology

The case of display systems presents an interesting example of a technology that is already forging a new path for electronic polymers. It is not surprising that con-

*Active conductive polymers find application in three areas of the LCD: the *integrated circuit,* which drives the display; the *glue or adhesive* used to attach metal to glass and which is likely to replace solder; and conducting channels, created through etching process (via lithography) technology, and which form long and narrow intersecting rows and columns ("row electrodes" and "column electrodes") to shape the display matrix.

TABLE 2.4. Markets for conducting polymers

Polythiopene	Antistatic treatment of photographic film
Polypyrrole	Battery electrodes
	Conductive coatings for electrostatic speakers
Polyaniline (doped)	Battery electrodes
	Conductive coatings for electrostatic speakers
	Antistatic materials in plastic carpets
	Corrosion inhibitor
	Shields to block radiation from electronic equipment
Polydialkylfluorenes	Video and TV color screens
Polyphenylenevinylene (PPV)	Electronic displays
	Radio-frequency identification equipment (RFID)
	Sensors
	Plasma and liquid crystal displays (LCDs)
	Solar cells
	Light-emitting technology

ductive polymers first entered commercial application in display technology application. We need only consider the critical market drivers. First, there is optical integrity and variety. By manipulating the molecular structure of OPEs, their optical properties can be custom designed to meet the specific needs of a particular display-using device. By varying the chemical composition of electronic polymers, these organic materials provide a continuum of emission and illumination options across the visible spectrum for a wide range of applications. Light of any wavelength can be created for wide color possibilities. Thus, the view of the display is excellent from any angle (i.e., no distortion). This benefit avoids the streaking (or ghosting) of images that is associated with LCDs (because the small plastic light emitters switch on and off very rapidly in contrast to liquid crystal materials, which do not respond rapidly enough to images when they flash across the screen).

Second, as displays find their way into more complex electronics, they must meet heightened data-processing speed requirements. Advancing electronic polymer material–dopant combinations, as well as more precise circuit-printing technology, impart to organic polymer compounds increased conductivity and, in turn, accelerated data retrieval, memory, and processing capability.

Third, display technology must be able to operate under reduced power requirements. In portable devices, lower power means longer battery life. To a greater degree than silicon, conductive polymers can be custom designed to operate displays under low-power conditions. Finally, there is the criterion of flexibility, a crucial property of advanced display systems; electronic polymers are by nature mechanically flexible.

Organic polymer displays, in the form of organic light-emitting devices (OLEDs), entered the market in 2000, and by early 2003 had already captured a United States market of nearly $100 million. There have been important advances in augmenting efficiencies of OLED light emissions, which has led to their growing use in battery-powered electronic devices, such as personal digital assistants. Improved processing has also led to declining production costs.*

Assuming that reliability, stability, and flexibility of OPEs continue to advance, display-using devices will increasingly incorporate OLEDs, including displays used in toy products, hand-held calculators, and "touch-screen" applications (see Table 2.5).

OPEs and the Integrated Circuit

As important as optical applications may be, champions of OPE materials have their sights on even bigger game—the integrated circuit itself. In the mid- to long-term, electronic polymers are poised to replace silicon itself in integrated circuits and electronics as a whole. Electronic polymers, and, in particular, the PPVs, have potential competitive advantages over silicon. Most importantly, compared to silicon, there is virtually an infinite variety of OPEs (e.g., PPVs can be linked with a broad range of aliphatic organics and can be altered through the injection of dopants). Thus, OPEs are much more able to meet the particular needs of technologies and customers, a characteristic which will lead to commercial applications. Silicon-based technology, in contrast, is a much more static material (i.e., less open to change and adaptation to shifting market demand).

The advantages of OPEs are not just limited to extent of application; there is a cost element to consider as well. Simply put, silicon-based technology has always entailed considerable expense. The equipment required by the vacuum deposition process for silicon is significantly more expensive than the equipment used for the spinning process of the OPE technology. For instance, costs of vacuum deposition equipment can approach $1,000,000, whereas equipment for the spinning of OPE may involve expense of only a few thousand dollars. Also, the need to employ vacuum methods in a highly sterile environment boosts costs of silicon-based production many-fold.

Advanced (Nonthin) Coatings [8]

Advanced coating materials are related to, but certainly distinct from, the polymeric substances. They comprise both organic and inorganic components. The new coatings protect surfaces from the environment—heat, impacts, erosion, and chemical degradation—and heighten the ability to sense and respond to the full range of

*In one process, small-molecule OLEDs are grown on a polymeric substrate coated with a transparent conducting substance (e.g., indium tin oxide or polyaniline) that is a de facto anode. This procedure results in a multilayer substrate about 100 nm thick. A second procedure involves deposition of another layer made of a "hole-transporting" organic substance. A cathode is then deposited, composed of a metal with a low work function, thus assuring a very efficient, low-resisitance injection of electrons from the cathodes, such as a magnesium–silver alloy.

TABLE 2.5. Probable display markets for conducting polymers

Year	Market
2010	Toy products
	Hand-held calculators
	Touch-screen applications
	Large-screen TVs
	Automotive displays
2015	Cellular phones
	Light-emitting billboards
	Automotive displays
	Digital cameras
2020	Smart phones
	Automobile navigation systems
	Notebook computers
	Internet appliances

changes that occur within the surroundings. Advanced coatings further expand the envelope of new materials technology through their essential role in so-called "smart" materials. These new coating materials act as thermal barriers, conductors, or anticorrosion coatings, or, in their most sophisticated form, provide multifunctional technology.

Thermal Barrier Coatings

Thermal barrier coatings protect surfaces in one of three ways: providing a simple physical barrier to thermal energy ("passive" heat control); dissipating, dispersing, or reflecting heat; or minimizing heat-producing friction. A new generation of thermal barrier coatings looks to protect surfaces from very high temperature environments by affecting all three modes of heat management.

Thermal coating materials include a variety of aluminum alloys and metal-matrix composites, ceramic-based materials, aluminum oxides, titanium alloys, zirconia–yttria compounds, and molybdenum plasmas. One of the more promising passive thermal coating materials emerged from advances made in polyimide chemistry. Polyimide coatings withstand temperatures of up to 700°C, or approximately 50% higher than current coating materials. The polyimide materials, made in thermal reactors under relatively low pressures, also impact surfaces with superior resistance to corrosive agents.

The markets for advanced thermal coatings range from the automotive, aerospace, and defense industries to high-temperature microelectronic circuit boards, industrial motors, electric power generation (including nuclear), biomedical systems, chemical and petrochemical plants, and composite materials for construction applications and machine tools.

Conductive Coatings

Conductive coatings consist of an electrically conductive material mixed into, or bonded onto, some nonconductive medium through such means as vapor-phase deposition or electroplating. In this sense, conductive coatings fall into the category of composite materials. Currently, conductive coatings exist commercially in three main forms, defined by the type of medium employed: conductive paint, metal plating (or cladding), and synthetic resin (e.g., epoxy, urethane, or acrylic)

The new conductive coatings, both metal and polymer, find application in electrical and opto-electronic systems. Conducting coatings, when incorporated into a battery's current collector, enhance the power and life of the battery. The coatings impart portability, compactness, and lower costs as well. Adhesives made from conducting coatings (e.g., epoxy medium) can repair printed circuits and replace metallic solder. Conductive coatings on glass substrates are the central technology in two-dimensional antenna systems for use in automobiles and telecommunications equipment. Transparent conducting films find application in optical systems, dielectric mirrors, and holographic devices.

One of the most important applications for metal-based conductive coatings is in electromagnetic interference (EMI) shielding. Conductive coatings absorb, emit, or reflect certain optical and radio frequencies. Moreover, they create a magnetic field or three-dimensional geometry that scatters radar signals, reducing the signatures of aircraft and ships. The coating protects equipment from interfering signals and sudden and potentially disruptive electromagnetic pulses. Conductive coatings can shield entire rooms containing electronic equipment or replace plastic as the packaging material for printed circuits and electronic components and devices (e.g., computers and mobile phones). Industries that employ conductive coatings in shielding systems include the aerospace, defense, electronics, security, health care, financial, and communications sectors.

Anticorrosion Metallic Coatings

Significant problems continue to plague current anticorrosion coating technology. Currently used coatings carry with them environmental problems and require expensive and time-consuming preparation of the surface. Also, coatings degrade over time, resulting in flaking and peeling. Promising materials under investigation are non-solvent-based coatings incorporating advanced polymer materials including polyester, polyaniline, and silicone and silicon–glycol resins.

These coating systems severely challenge current technologies. A potentially revolutionary line of research being conducted in the United States and Europe concentrates on organic films that form tightly bound multilayers on a surface through electrolytic action. These materials constitute gel-like films of alternating layers of positively and negatively charged molecules. As opposing charges pair up, they hold adjacent layers together tightly while, at the same time, a positively charged bottom electrolyte layer adheres to a negatively charged metal surface, thus avoiding degradation and flaking over time.

Not to be outdone by these organic materials, purely metallic (that is, inorganic) anticorrosion coatings involve complexes of aluminum, a rare earth metal (e.g., cerium), and a transition metal (e.g., iron or cobalt) combined in various proportions. The nature of the alloy itself, produced by an innovative thermal process, allows quenching of the molten metal at a relatively low rate (as measured in degrees cooled per second) compared to current aluminum alloy materials. This less radical quenching process, undertaken in a thermal furnace, produces an amorphous alloy without structural damage to the metal. This noncrystalline structure serves well as both an anticorrosion and antideformation coating.

Commercialization efforts are proceeding on totally new processes for making thin films, such as the process based on the creation of ionic self-assembled coating layers. In this approach, a charged substrate is dipped into an aqueous solution of a cationic material (i.e., positively charged), followed by a second dipping in an anionic solution (i.e., negatively charged). Adsorption to the surface of the substrate results from electrostatic attraction of "interlayer charges," with each layer of uniform thickness. Multilayers several microns thick are easily fabricated through repeated dipping processes and are rapidly dried and fixed at room temperatures. The low-cost process produces an ultrathin, impermeable, and tightly bound coating. The process produces highly specific coatings, depending on the applications involved, through molecular manipulation.

Continued innovation and the decreasing costs over time in making the coating material and in applying the coating to a surface are expected to expand markets for the technology. In addition to making further inroads into such traditional markets as shipbuilding and repair, public infrastructure (bridges, buildings, etc.), public utilities, machinery, buildings, and construction, the technology will find increased applications in such industries as aerospace, automotive, electronics, industrial gases, telecommunications, and petrochemicals.

Multifunctional ("Smart") Coatings

Multifunctional coating technology leads us to the pinnacle of capability within the coatings field as a whole. Multifunctional coatings, which first emerged in the 1990s, perform a number of operations—anticorrosion protection, conduction, electromagnetic shielding, and thermal protection—simultaneously and in an interactive manner. Their development, both within the United States and internationally, results from exploiting the commercial potential of surface engineering. Typically, small R&D and start-up firms license multifunctional coating technology from the government and universities. In addition, certain large corporations (e.g., Dow Corning) look to expanding their product capability in the field.

Multifunctional coatings incorporate new materials, either separately or in combination. These materials include fluoropolymer composites, the urea–formaldehyde resins, multicomponent pigments, and the carbides, nitrides, and borides of certain metals (e.g., titanium). These materials can disperse within different media such as paint, ink, and adhesives. Multifunctional coatings may be composed of one material capable of performing different functions, or, more commonly, a multilayered composite of different materials, each performing a single but related task. Increasingly, work in the field focuses on the synthesis of nanocomposite coatings with multifunctional properties that create a full spectrum of surface qualities, including transparency, surface hardness, reflectivity, and so forth.

Numerous applications exist for multifunctional coatings. In the automotive sector, multifunctional coatings meet increased demands for strength and thermal and corrosion protection of steel surfaces. In the textile industry, research in the United States and Germany looks to development of "smart" hybrid polymeric coatings for fibers that allow fibers and textiles to adjust or tune their properties in response to external stimuli. In the metallurgical industries, multifunctional coatings providing superior hardness, anticorrosion properties, thermal protection, abrasion resistance, and chemical "inertness," enter into complex metallurgical operations, such as pressure die casting processes. The defense and aerospace industries use multifunctional coatings in complex information systems technology as well as aircraft for sensing, conductivity, energy absorption, and thermal dissipation. Such coatings increase the performance capability and lifetime of components, equipment, and defense systems.

An emerging application for multifunctional coatings is in a new generation of micro-electromechanical devices (MEMS) requiring the simultaneous detection of temperature, pressure, radiation, gas concentrations, and electromagnetic fields, in multicomponent mechanical systems. Multifunctional coatings also act as sensors to detect and monitor structural defects in buildings, bridges, and aircraft, and to carry and deliver chemical agents to strengthen critical points in mechanical structures. In one variation of the technology, the sensing system uses small synthetic spheres arrayed in a crystalline lattice and embedded within a coating material. As the coating shifts or otherwise changes its configuration due to structural distortion, the internal lattice also changes its structure. An optical system then monitors these changes over time. These spheres also contain various anticorrosion agents and deliver them to pivotal sites in a structure.

Nanopowders and Nanocomposites [9]

These materials have made enormous technical and commercial progress starting in the late 1980s. Nanopowders evolved from the powdered metal field, which itself is a growing presence in industry. Nanopowders are typically metals or metal mixtures and complexes with particulate sizes in the micron ranges. The market application of these materials depends on the fact that they can be formed into diverse shapes and forms possessing unique and useful mechanical, electrical, and chemical characteristics.*

*The metals and metal complexes most closely associated with nanopowders include the oxides of aluminum, magnesium, iron, zinc, cerium, silver, titanium, yttrium, vanadium, manganese, and lithium; the carbides and nitrides of such metals as tungsten and silicon; and metal mixtures, such as lithium/titanium, lithium manganese, silver/zinc, copper/tungsten, indium/tin, antimony/tin, and lithium/vanadium.

Nanopowders demonstrate the interrelationships that occur within the different sectors of the advanced-materials field since these materials intersect intimately with both coatings and polymers. Nanopowder coatings possess more tightly packed structures than exist in the case of other coating materials. This structure, in turn, imparts to the surface a high degree of transparency, hardness, and abrasion and scuff resistance. These materials, when added to a resin base, produce superior paints and varnishes. Additional applications for nanopowder coatings include their use as abrasives for polishing silicon wafers and chips, hard disc drives (for higher data storage capability), and optical and fiber-optical systems; as advanced catalyst for petroleum refining and petrochemicals production, as well as in automotive catalytic converters (providing more complete conversion of fuel to nontoxic gases); as pigments in paints and coatings; and as additives to plastics in a new generation of semiconductor packaging.

Nanopowder technology is also used in the manufacture of nanopowder–plastic composites. Typically, these composites contain under 6% by weight of nanometer-sized mineral particles embedded in resins. One of the first such composites used nylon as the plastic medium. More recently, other plastics have come to the fore, such as polypropylene and polyester resins. Nanocomposites have such beneficial properties as great strength and durability, shock resistance, electrical conductivity, thermal protection, gas impermeability, and flame retardancy. New and more sophisticated processes use the delicate connectiveness between nanocoatings and plastics through the manufacture of composite powders with a uniform, nanolayer-thick metallic or ceramic coating for high-density durable goods parts.

Due to their superior properties and advancing manufacturing technology, nanocomposites have led to numerous market opportunities. Indeed, both nanopowders and nanocomposites diffuse throughout industry at a rapid rate. Their range of application currently outpaces the assumed "stars" of the advanced-materials group, such as the nanocarbon materials. To capture a sense of this diversity and market power, we need only mention the different industries that employ these materials as powerful substitutes for the better known metals and plastics. For example, in the automotive area, General Motors recently began production of the first polymer nanocomposite part for the exterior of a car. The biomedical field also is adopting nanopowder composites, especially as delivery systems for the application of bioactive agents into the body, as materials for dental and medical microabrasion applications, and for use in orthopedic implants (e.g., artificial bones and hips) and heart valves. Table 2.6 displays the growing spectrum of application for nanopowders and nanocomposites.

Nanocarbon Materials [10]

It is fair to say that nanocarbon materials have received more attention and press space than their actual achievements to date warrant. The great commercial surge of these materials is yet to happen. This being said, these materials do hold great

TABLE 2.6. Current and emerging applications for nanopowders and nanocomposites

- Electrodes for portable power sources (batteries and solar cells)
- Military weapons (e.g., as armor and in projectiles)
- Advanced instrumentation (e.g., for automotive applications)
- Biomedical and environmental sensors
- Stronger, lighter, and more flexible structural shapes
- High-performance cutting tools and industrial abrasives
- Advanced refractory materials for chemical, metallurgical, and power generation
- Ceramic liners (made of zirconia and alumina) in more efficient internal combustion engine cylinders and ignition systems for automotive and aerospace applications
- Industrial magnets in magnetic resonance imaging (MRI) systems for medical applications
- New generation of electrical and electronic components (e.g., induction coils, piezoelectric crystals, oscillators)

promise for the future and by the first decade of the twenty-first century had clearly begun to demonstrate their practical mettle in the workaday economy. These intriguing and diverse substances (in a manner similar to the advanced polymers) combine in a symbiotic matrix carbon (organic) and metal (inorganic) nanosized particles arranged in carefully designed spatial configurations. One such group falls into the category of fullerenes. In this case, a series of spherically structured carbon atoms enclose one or more metal atoms. In the second type of material, the carbon atoms join together to form a tubular-like structure, which may or may not engross metal ions. These so-called "nanotube" materials have been gaining significant ground in advanced composite applications.

Metal Fullerenes

Fullerenes in general refer to a group of materials composed of carbon structures of 60 to 90 carbon atoms, each enveloping a single metal atom. At the end of the last century, a number of U.S. companies manufactured fullerene materials in varying compositions and amounts. Since the 1990s, research undertaken in the United States has led to the creation of the "tri-atomic" fullerene, in which the carbon cage contains three distinct metallic atoms. These materials possess commercially useful properties now being explored by research and industry groups.

These advanced fullerenes offer a variety of potentially important applications. For example, they are at the heart of new types of multifunctional catalyst systems for the petrochemical industry. In this context, the carbon structures encapsulate the different catalytically active metals (e.g., iron, platinum, and nickel), which are then released in tandem and in a controlled way as the external carbon structure disintegrates during reaction. The unique optical properties of the fullerene materials offer additional applications in industrial photovoltaic sensing systems for incorporation into monitoring and automated-control technology. The electromagnetic properties of these materials also apply to semiconductor, fiber optic, and microelectronic systems, although these uses remain to be exploited commercially.

Within the biomedical field, advanced fullerenes act as superior "contrasting" agents for use in magnetic resonance imaging (MRI) systems. In this case, the fullerenes, ingested into the body orally, enhance MRI images 50 to 100 times more than previously available agents. As a consequence of this improved MRI performance, manufacturers can incorporate smaller and less powerful magnets in their machines, resulting in more compact, portable, and cheaper equipment. This advantage, in turn, potentially expands the markets for MRI technology into rural, less developed regions, smaller to mid-sized clinics and hospitals, and military field hospitals. Further, the smaller MRI devices, because they operate with less powerful magnetic fields, reduce the costs (as well as potential dangers) to the larger hospitals and clinics of housing and maintaining large superconducting magnets.

Nanotubes

Nanotubes are close cousins to the metal fullerene group. At the same time, they are distinct structures with specific applications. In essence, nanotubes are carbon-based structures with cylindrical shapes and diameters between 0.8 to 300 nanometers. Nanotubes resemble small, rolled tubes of graphite. As such they possess high tensile strength and can act as an excellent conductor or semiconductor material. There are two main varieties of nanotubes: single-walled and multiwalled. Multiwalled structures represent the less pure form of nanotubes and offer only a limited number of applications. The more advanced, purer form of nanotube, defined by a single-walled structure, is the more promising material commercially, especially for incorporation into polymeric materials in the synthesis of composites with superior structural, thermal, and electrical characteristics.

Although in 2006 there were about twenty nanotube producers worldwide, these materials had not at that time reached the mass commercial stage. Within the United States and worldwide, nanotube sales remain small. Within the United States in 2005, the plastics industry bought about $5 million worth of nanotubes. In that same year, advanced composites accounted for only $4 million in sales, with the fibers and textiles industries buying another $3.5 million, and research groups around $1 million. Other countries looking to mass produce the material are Japan, China, South Korea, and France. The general expectation is that the market for nantubes worldwide will accelerate after 2010. It is expected that the most likely near-term markets are catalytic converters for the automotive industry, lightweight and durable materials for the aerospace industry, and stronger armor and field suits for the defense sector.

As with the metal fullerenes, and to an even greater extent, large-scale markets for nanotubes exploit this material's electronic and optical applications. Because nanotubes have dimensions in the wavelength range of visible light, they can be used directly as active opto-electronic devices. For example, Motorola, Samsung, and other electronics companies have been developing advanced electronic displays based on nanotubes. This work is leading to ultrathin screens and flat-panel displays capable of high-resolution imaging and superior power efficiency, and to a new generation of large-display, low-cost illumination systems. A potential market for nanotube display technology is for 20 to 40 inch television screens since neither LCDs nor other existing display technologies have as yet secured a dominant position in the field. However, organic electronic materials, as previously discussed, may also soon enter the large-screen markets through OLED technology. In this case, a heated competition for this market between these two emerging advanced materials will likely take place.

A related area of interest is the use of nanotubes in microelectronic devices. In particular, IBM recently succeeded in making microelectronic switches from nanotubes. This device has applications in computer and consumer electronic products. The aerospace and defense industries promise markets for advanced nanotube composites as well. These composite materials are both strong and light (20–30% lighter than carbon fibers) and, consequently, make excellent materials for aircraft components and structures. Additional potential applications for carbon nanotubes include incorporation into thermally conductive fibers for clothing, carpets, and fabrics; electrically conducting polymers and fibers for use as electromagnetic shielding materials; and various components for wireless communications, microsensors, and monitoring devices. Over the longer term, nanotube composites are likely to provide superior drug delivery systems as well as advanced storage systems for hydrogen-based fuel cells.

Nanofibers [11]

Nanofiber technology refers to the synthesis by various means of fiber materials with diameters less than 100 nanometers. Nanofibers depend on their high flexibility and, therefore, their ability to conform to a large number of three-dimensional configurations. They also have a very high surface area, offering a myriad of possible interactions with chemical and physical environments. Recent research suggests possible industrial applications as ceramic ultrafilters, gas separator membranes, electronic substrates, medical and dental composites, fiber-reinforced plastics, electrical and thermal insulation, structural aerospace materials, and catalyst substrates for petrochemical synthesis. Nanofibers also may be applied in advanced optical systems, according to the shape, number, and composition of the fibers. As with the first-generation synthetic fibers of the mid-twentieth century, the most promising applications for nanofiber materials is in new types of textiles. Nanofibers potentially can impart beneficial properties to both natural and synthetic fibers, such as superior thermal insulation, durability, strength, resilience, texture, wrinkle resistance, and flexibility. Nanofibers may become the fabric fiber itself through polymerization or, in the form of ultrathin whiskers, be added to a traditional fiber to modify its properties.

Nanofiber research and development is highly active in the United States and Asia, especially South Korea. Within the United States, the two leading players in the field are large, well-established corporations with storied histories in the textile field. One of these is Burlington Industries (Burlington, NC) in its partnership with Nano-Tex (Greensboro, NC). DuPont, that monument in the field of synthetic fibers, reasonably looks to mine its expertise in macromolecular synthesis. For example, DuPont pins its hopes in the field on polymerizing textile-grade nano-sized fibers and by deepening its activity in a type of biosynthesis that effects various polymeric combinations of protein materials. In another approach, the company is pioneering the so-called "phase change materials (PCMs)." These textile materials consist of a selected fabric coated with synthetic materials encapsulated in plastic-based spheres. These embedded particles respond to changing body temperatures in a cyclic manner. As body heat increases, the PCM materials melt, thus drawing heat away from, and ultimately cooling, the body. As the body cools down, the PCM freezes again, in turn releasing stored heat for warmth.

A third player in the field is the U.S. government, specifically the Army allied with the Massachusetts Institute of Technology's Institute for Soldier Nanotechnologies. The textile materials being explored are designed to have both defense-related as well as commercial applications. This alliance pursues novel R&D paths such as water- and germ-proof nanoparticle coatings; super-strong bullet-proof vests made of fiber–nanotube composites; so-called "dynamic armor" that can detect the sound of bullets and "firm up" in response to repel the projectiles; hollow fibers embedded with nanoscale magnetic particles that can stiffen to transform into an instant splint for use in battlefield injuries; "exomuscle" uniforms providing soldiers with super-strength camouflage fabric made of high-performance microfibers.

The marshaling of such large amount of financial, technical, and human resources that the large corporations and government agencies are able to muster and exploit must seem daunting to the small start-ups that increasingly dot the advanced-material landscape, especially within the United States. But these micro- and small-to-medium sized firms appear capable to think creatively and move quickly on new ides and market opportunities. Such is the case in particular with nanofibers. Just as with the metal fullerenes, these flexible and innovative small- and medium-sized enterprises (SMEs) lead the way in the application of nanofibers in the biomedical area, especially in the integration of these new materials in the next-generation drug delivery systems and in advanced surgical-site structures and tissue-growing systems. These SMEs generally license such processes from universities (or government) in order to bring the relevant patents into the marketplace. Nanofibers being developed by SMEs are designed to produce three-dimensional collagen-based matrices or "scaffolds." When these scaffolds are "seeded" with specific types of human cells, blood vessels of small diameter form relatively rapidly. These vessels can then be transplanted into a patient. This application of nanofiber technology offers one of the most promising routes to man-made blood vessels. Because the fibers closely resemble naturally occurring tissue, cells readily grow in the man-made scaffold. Over time, this new material technology will catalyze and guide the synthesis of organs, nerves, muscles, and other tissues. In the near term, the synthetic nanofiber collagen mats transform into an effective high-technology bandage to stop bleeding during surgeries and to act as scaffolding in order to speed growth of new tissue at the wound site.

Thin Films [12]

We witnessed in our overview of organic polymer electronic materials the importance of thin-film layering (such as in the making of OLED devices). Thin-film technology is one of the main fields of the new-materials revolution. Advanced thin-film materials represent one of the newest and most promising emerging material technologies. A clear distinction exists between the advanced coating materials and thin films. The former applies to traditional surfaces and these coatings have widths or thicknesses in the macro region. Thin films involve materials different from those used in coatings. They include polymers, metals, and polycrystals. Moreover, these are layered only a few tenths of an angstrom deep onto a foundation or substrate, such as glass, acrylic, steel, ceramics, silica, and plastics. Whereas coatings are applied to surfaces, thin films, when multilayered, are themselves devices or parts of devices at the operating center in a variety of products and systems, including consumer electronics and electronic components, telecommunications devices, optical systems (e.g., reflective, antireflective, polarizing, and beam-splitter coatings), biomedical technology, sensor systems, electromagnetic and microwave systems, and energy sources and products (e.g., batteries, photovoltaic cells).

New developments and commercial possibilities (and realities) abound in this rather complex field. The 3M Corporation, for example, is developing a new generation of fluoro-acrylate thin-film polymers that possess both electronic and anticorrosive properties. These polymers form ultrathin transparent coatings on a number of substrates, including copper, aluminum, ceramic, steel, tin, and glass. Possible applications for these types of materials include their use in wireless telecommunications systems, liquid crystal and electrochromatic display technology, reflective or light-emitting ("smart") windows, advanced sensors, magnetic and laser devices, piezoelectric products and systems, biomedical devices and implants, antistatic electronic packaging (e.g., for use in packaging and protecting microchips), photovoltaic systems, corrosion protection, and xerographic applications.

But, as is the case in this latest advanced-materials revolution, the organics no longer have the run of modern technology. The inorganics resurface here as a force to be reckoned with. Thus, inorganic, metal-based thin films are closing in on reducing the size of circuits and circuit components for electronic applications and may, in fact, compete against the polymer thin films themselves in these markets. Metal-based thin films claim superior purity and, therefore, durable interconnections between microcircuit components. These advantages point to future computers that are much smaller and operate faster than current technology. Metal-based thin films clearly point the way to a new generation of microelectronic and electromagnetic components, including capacitors, resisters, thermistors, transducers, inductors, and related elements. Specific types of metals, metal compounds, and alloys used in advanced thin films involve alumina, tantalum, nickel, nickel–aluminum alloys, copper, silver, silver–palladium alloys, platinum, and zinc.

Both polymer- and metal-based thin films also chart out the route to "printed" low-cost antennas for attachment onto different surfaces. These antennas possess large surface areas for capacitive coupling and may compete against certain types of metallic conductive coatings. Additional potential markets for thin-film materials include applications in more efficient photovoltaic systems, thermally and electrically conductive adhesives (for chip-to-substrate bonding or for connecting materials in electronic enclosures), thin-film transistors, carpets and fabrics, wireless identification tags, electrodes for ultrasmall electronic devices (replacing traditional materials such as indium tin oxide), future fuel cell components, and less expensive, smaller, and more advanced microelectromechanical systems (MEMS) that combine computers with tiny mechanical devices such as sensors, valves, gears, mirrors, and actuators embedded in semiconductor chips.

Advanced Composites [13]

Composite materials actively engage global markets today. New developments in the field continue to flow so that now we can define an advanced composites field. Generally speaking, a composite material consists of two or more physically and/or chemically distinct, suitably arranged or distributed phases, with interfaces separating them. Most commonly, composites have a bulk phase (which is continuous) called the "matrix," and one dispersed, noncontinuous phase called the "reinforcement," which is usually harder and stronger. In general, the bulk phase accepts the load over a large surface, and transfers it to the reinforcement, which, being stiffer, increases the strength of the composite. Bulk materials tend to be organic materials, such as Kevlar and polyethylene, although ceramics also serve in this capacity. The inorganics dominate as reinforcement materials. These include glass, metals, ceramics, and a host of elements and their compounds. Such materials are often fibrous (whiskers, sheets, etc.) and are made of glass, ceramics, alumina, and silicon carbide. Composites generally have the twin virtues of being both light and strong.

The first engineered composite was fiberglass, developed in the 1930s. Fiberglass consists of glass fibers embedded in a polymer matrix. It was not until the post-World War II period that these composites made substantial inroads in the market, finding application in industrial and consumer products. Its first major market came in the 1950s, in automobiles (notably the bumper section). The Cold War was a driving force in the development of composites for military uses and aerospace and rocketry (Sputnik). Rockets in particular needed composites to withstand shock and heat. At this time, fiberglass composites entered significantly into aircraft design. From 1950s to 1980s, the percentage (by volume) of composites used in airliners increased from 2% to over 10%.

By the 1970s, the first truly advanced composites came commercially available with the development of high-modulus whisker and filament-based composite materials followed by the commercialization of fiber-reinforced metal–matrix composites (MMCs). These were first used in the U.S. Space Shuttle project. A few years later came development of boron–aluminum composites with well-known business and consumer applications.

The larger companies tend to dominate in the manufacture of these older (pre-1970s) composites, the largest of these being Owens-Corning, which specializes in glass-reinforced composites. In recent years, more advanced types of composites have come to the fore that encompass a wider variety of both bulk matrix and reinforcements. The smaller- to medium-sized firms, often licensing technology from academic and government laboratories, are more often involved in bringing these newer composites into the market.

Two types of advanced composites currently being developed are nanocomposites and the so-called "smart" composites. Nanocomposites consist of some bulk matrix that engrosses nanosized particles of reinforced material. The small size of the internal additives imparts increased strength to material as a whole. Nanocomposites find significant applications in stronger but lighter-weight automotive parts. With their enhanced gas-barrier properties, they are useful in packaging and in improved flame-retarding systems.

Smart composites mimic certain characteristics of living organisms. These materials, with built-in sensors and actuators, react to their external environment by bringing on the desired response. This is accomplished by linking the mechanical, electrical, and magnetic properties of these materials through the incorporation of piezoelectric reinforcements. For example, an electrical current generated during a mechanical vibration could be detected, amplified, and sent back through the composite material, causing the latter to "stiffen" and cushion the vibrating components. This will have important applications in construction, such as to minimize structural damage to buildings during earthquakes.

The development of new and advanced composites goes hand-in-hand with innovation in processing. The manufacture of composites can take place in the solid, liquid, or gaseous phases. Solid-phase methods include powder metallurgy and foil-diffusion bonding. In the first, powdered metal and discontinuous reinforcement material are mixed and then bonded through compacting, degassing, and thermomechanical treatment. In foil-diffusion bonding, layers of metal foil are sandwiched with long fibers, and then pressed together to form a matrix. Liquid-phase processes include electroplating, by which a solution containing metal ions loaded with reinforcing particles is codeposited, forming the composite material. Other processes include squeeze casting, involving molten metal injection into a form with fibers preplaced inside; spray deposition, in which molten metal is sprayed onto a continuous fiber substrate;, and reactive processing that requires a chemical reaction between matrix and reinforcement materials. Finally, an important gas-phase method is physical vapor deposition (PVD) in which a fiber reinforcement passes through a thick cloud of vaporized metal. These processes are leading to advances in efficiencies and product quality.

GLOBAL MARKETS: THE QUESTION OF CONVERGENCE

The range of application of advanced materials clearly extends across many industries. Technological change within these industries, and thus national economies overall, increasingly depend on the further development and commercialization of these advanced-materials products and the processes that create them. As an advanced-material technology proceeds along its product life cycle, its market grows in volume and diversity. Market volume expands as unit price contracts—due to learning curve effects—and the technical capabilities of the product extends its reach. Market diversity then broadens as researchers learn how to modify the material in order to take advantage of different commercial applications.

This is not to deny the many and daunting risks involved in cosseting a new material from R&D, preproduction, commercialization, and entrance into the final market. We will have more to say on this in following chapter but, for sake of argument, assuming that companies, industries, and countries manage their technologies effectively, these active players can reduce, even minimize, the risks inherent in the introduction of such new technology and so capture a sizeable portion of the growing markets. And indeed, there is evidence of this already. In the United States alone, advanced materials have begun to proliferate. We estimate that one-quarter of all advanced-materials companies within the United States already have products on the market. Although annual revenues for these companies remain relatively low—one-half of these companies claim annual sales revenues of less than $15 million—the size of these firms are on the rise. Whereas, in 1990, only a very few firms had annual sales greater than $50 million, today we estimate that one-fifth of advanced material firms exceed this amount [14].

What then can we determine as to the present and future for these materials? Our concern here is not just in the global sense, but, within that broad canvas, the varying rates of growth in demand between the United States and Europe and, more particularly, the European Union. Economists and economic historians often point to the large and growing market within the United States through the nineteenth century as the pivotal factor in the rise of America's most important innovations. With such a (more or less) homogeneous market, the "American System" flourished and evolved into the full glory of mass production, that technological and organizational juggernaut that catapulted the United States into the leadership role of the industrial revolution. Even if we accept this market-driven paradigm of progress and growth as applied to the past, are we then to conclude that this dynamic continues to be the determinant of the rate and direction of 21st century economies? In fact, while not singling out demand as the sole explanation for economic expansion today, theorists of national competitive advantage—most notably Michael Porter—maintain that it remains a vital component in explaining differences in technical and economic performance between regions and countries.

If this is so, how are we to plumb the past and present market for what we have termed "advanced materials," and more daunting still, how can we reasonably forecast what this market will be ten or twenty years hence, not only globally but disaggregated to the national and regional level? Certainly, those who begrudge us even the possibility of being able to clearly delimit the advanced-materials industry itself, proclaiming that it is too widely diverse and inchoate a group, must of necessity, deny us even the remote possibility of even beginning such a market analysis. But we have seen from our descriptive overview of the major advanced-material groups that a definite industry can be discerned; the new generation of advanced materials can indeed be clearly defined. As we have seen, these materials are all linked—"red-lined" if you will—by a common generalized "process" that discounts macromolecular synthesis in favor of the imposition, manipulation, and reconfiguration of atoms and atomic groups within smaller than macro molecular units. Through this understanding, entirely new types of materials emerge with a greater flexibility of application and range of usefulness than has been true with past material systems. It is largely for this reason that we can today equate new materials with new technology in general to an extent never before possible.

Nor do we, in some Cartesian way, have to deduce from our common process what these materials are, a method far too theoretical and far too open to questions, disputation, and counter-argument. We undertake a far more empirical approach in identification of the advanced-materials group. A traversal of the volumes of articles, both scholarly and trade, as well as business plans of large corporations on one end of the size spectrum, and start-ups on the other, and government reports from the United States and Europe on advancing technology, tell us empirically and absolutely what we need to know, with evidence that cannot be so easily challenged. These sources clearly lay out the most important new material technologies, those already on the market as well as those in development and expected to be brought to commercial application in the near term. These technologies, products, and product groups clearly define the boundaries of the advanced-material movement today to the extent that we know what is part of, and not a part of, that privileged collection of modern-day innovations. For example, we can easily discard such macromolecular polymers as nylon, polyester, and so forth as well-known, impacted technology. On the other hand, organic polymer electronic materials need to be included in our industry.

Through this process of constructing that critical boundary that delimits the group of advanced materials and, by the same token, excludes those of less interest, we construct a clear understanding of the extent and limits of this new-materials revolution. This is all fine and good as long as we remain within the realm of the qualitative. But what recourse do we have if we need to understand trend lines in demand worldwide as well as within different countries? We could, of course, take past trends in these materials as our cue and extrapolate from there, but this begs the question of where exactly we will locate such data. Traditional sources, such as the U.S. Department of Commerce (and counterpart government agencies in other countries) have limited use. Their statistics do not include our new materials for the simple reason that they are too new and unfamiliar and do not have a long enough track record to enter the radar screen of those that compile industry facts and figures. Business plans created by new firms and start-ups for investor groups would seem to be good, solid sources, somewhere between primary and secondary in nature. But the sales projections proffered in these all-to-hopeful assessments do not inspire confidence in their reliability. These sales forecasts typically paint to the reader—that is to say, the potential investor with his money on the line—far too rosy a picture of things to come. The projected sales figures tend to be grossly inflated, although they do measure at the least the economic impact of these materials. But even here, these analyses do not generally take into consideration the impending and dangerous risk factors that can gut the most hopeful firm projections and that must be carefully weighed if any sort of reliable forecasts are to emerge.

We are most likely to be on safer ground if we take previous technologies as our models for growth. We know from innovations of the past century the life cycle profiles of new products and processes. These life cycle profiles measure the percentage of the total expected market within those industrial sectors that are (or will be) major consumers of these materials. The typical profile describes an emergent new technology with only a small percentage of the market. As the new technology becomes more familiar to the market and as improvements are made over time to the product (higher quality) and process (lower costs), diffusion accelerates until it reaches a maximum point of market saturation. With the entrance of upstart competitors, market penetration takes a downward turn from this peak. We should then consider this life cycle model as an approximate guide for estimating the market evolution of each of the advanced-material groups. In doing so, we adjust or perturb this bell curve model as seems necessary from the individual histories and tendencies particular to the various advanced-material categories. Thus we find that in some product life cycle profiles, the maximum or peak point can be delayed or may not show up at all. In this latter situation, the high point does not occur until past 2030, the last year for which we attempt a market forecast.

Variations from the expected bell curve life cycle profile occur for a number of reasons. Ultimately, the criterion is whether that product is anticipated to perform better or worse than the "norm." This assessment, in turn, depends on such factors as expected rate of market acceptance and "staying power" within the market. Such factors appear to be strengths and limitations inherent in the product itself and not dependent on how well the commercialization of a technology has been managed. In other words, the profiles considered here should be taken as the optimal that can possibly be hoped for by those stakeholders associated with them. For example, the literature signals to us that, even if developed and guided into the market by hyper-competent management teams, there is little chance that some materials, by dint of their very structures and characteristics, will be able to keep up the pace with other, better-performing materials. But we are not by any means advocating a form of technological determinism. How well these technologies, in fact, are managed will determine how close the material can come to matching these optimal performance profiles. It is the manager of technology then that has the final say. This last question relates to degree of risks and the management of these risks by companies and by countries as well. In the analysis to follow, we simply assume that management handles the new technologies optimally. The question of risks and how well they are really managed (especially in comparing the United States and the European Union) will be given due consideration in later chapters.

Table 2.7 displays (first row) the life-cycle profile of a typical (and hypothetical) innovation, assumed to have been introduced in 1970, through 2030. This technology might, for example, be a radically new synthetic fiber. The numbers represent market penetration. We note that the maximum point is reached in exact middle of the cycle, in the year 2000, after which decline sets in with market saturation and increased competition. The profiles that follow are believed to be the most likely to occur for the various advanced-material groups. They are constructed from case study analysis of each product. The market penetration figures shown indicate the percentage of the total market for that product group captured by the advanced material. For example, in the case of advanced ceramics, we assume the total market to be the ceramics market as a whole; for nanofibers, it is the fiber market as a whole; for advanced composites, it is the total composites market; and so forth. As is seen, the different advanced-material groups are not all in phase in their cycle. Some products began their ascent early on, whereas others, still under development, did

TABLE 2.7. Ideal life cycle patterns for the new materials: market penetration rates within their particular industries (%)

	1970	1975	1980	1985	1990	1995	2000	2005	2010	2015	2020	2025	2030
Typical profile model (for an innovation introduced in 1970)	5	10	25	35	50	65	70	65	60	50	40	35	30
Bioengineered materials	—	—	—	—	—	—	1	2	3	5	10	15	20
Advanced 2 ("super") alloys	2	3	4	5	6	7	10	12	15	20	25	30	
Advanced ceramics	—	—	1	2	3	5	10	15	20	25	30	35	40
Engineering polymers	3	5	8	10	25	35	40	45	50	50	48	45	40
Organic polymer electronics (OPEs)	—	—	—	—	—	—	1	3	5	10	25	35	40
Advanced electronic materials (other)	—	3	5	8	15	20	25	25	30	35	40	45	45
Advanced coatings	—	—	—	—	1	2	4	5	7	10	25	35	40
Nanopowders	—	—	—	—	1	3	5	8	10	15	20	30	35
Nanocarbon materials	—	—	—	—	—	—	1	2	3	8	12	20	30
Nanofibers	—	—	—	—	—	—	1	2	3	5	10	20	25
Thin films	—	—	—	—	1	3	5	8	12	15	18	24	28
Advanced composites	—	—	1	3	5	8	12	15	18	21	25	28	30

Source: [15].

(or will) not enter the market until later, at which time their market profile will begin its life cycle.

These market penetration percentages can then be applied to the total market associated with each product group based on estimates and (after 2005) projections of each advanced material's relevant market through 2030 over time and geographical region. For this analysis, our concern is with the United States and Europe. In this scenario, we assume a convergence model so that while different rates of growth must be assumed between countries and regions initially for each advanced material's relevant market, we make the assumption that these gaps between regions narrow significantly as the forces of globalization surge forward.

Within the United States, the markets relevant to advanced materials (e.g., fibers, plastics, composites, etc.) experienced more or less steady growth from the 1970s to 2000. Following the recession of 2000–2003 and the economic instabilities of 2007–2009, these industries can expect steady growth until 2020 and beyond.

In the wake of the September 11 and the wars in Afghanistan and Iraq, the Depart of Defense (DOD) also has become a major consumer of technologies that include the very latest in advanced materials. The government, especially the DOD, has its own unique and expanding demands. Advanced materials represent a growing percentage of this country's national security budget. Indeed, in 2003, DOD committed in excess of $200 million for research in projects related to advanced materials of various kinds. By 2015, we estimate that this figure will grow to $950 million. Defense applications of advanced materials include uniforms, weaponry, armaments, sensors, biomedical, computers and telecommunications, food, and other materiel concerns. [16]

On the other hand, a convergence-based model must assume greater competition, first from Europe and then from Asia. This means that the United States will experience an eventual contraction in the percentage of the relevant markets it controls.

If we assume that this is the best of all possible worlds and convergence is indeed the order of things, the European Union can be expected to eventually catch up with the United States in terms of proportion of global markets captured. Despite the constitutional set backs of 2005–2007, the European Union remains a growing economic force that could potentially offer a real challenge to the United States. the European Union's markets and currencies are now more unified than before and the addition of the ten Eastern European countries in 2004 greatly extended the human and material resources of the Union. As previously discussed, the European Union is committed to developing its high-technology capability. England, France, and Germany continue to conduct scientific research and development on cutting-edge materials and their applications. With trading barriers within Europe weakened and globalization on the rise, countries such as Ireland and Spain, and the nations of Scandinavia, show a similar urge to extend their industrial base. Ireland in particular enjoys a distinct success in the areas of electronics and information technology, and, increasingly, biotechnology. The eastern portion of the European Union remains a question mark, but an interesting opportunity as well. Most Eastern European countries undertake some sort of research and development work in advanced materials and offer cheap land and labor as well as favorable tax schedules that are expected to attract industry from the western parts of the European Union as well as other parts of the world

In our hypothetical convergence model, we see an expanding U.S. economy in the wake of technological progress, but one that must eventually make room for a Europe that has been growing and gaining momentum through the forces of globalization and, finally, an Asian presence that takes advantage of open western markets and, because of new technology, cheap modes of transportation and materials management.

From these assumptions, we calculate the global market for each advanced material category by: (1) projecting to 2030 the global market for each of the relevant sectors (e.g., total fiber market, total composites market, total ceramics market, etc.); and (2) multiplying this result by the percentages that apply to the life-cycle patterns for each advanced material category (see Table 2.7). We then determine the market for each advanced material category for the United States and Europe by: (1) projecting to 2030 the global market for each of the relevant sectors (e.g., total fiber market, total composites market, total ceramics market, etc.); (2) multiplying these total markets by the percentage of the market that we deem belongs to the United States and Europe at any particular time, given the assumption of the convergence model (i.e., sometime around 2010 and in the years following, this percentage declines for the United States as it faces increased competition from Europe and Asia, with Europe itself eventually facing declining percentages in the face of a competitive Asia); and (3) multiplying this result by the percentages that apply to the life-cycle patterns for each advanced material category (see Table 2.7), assuming these patterns are the same across regions.

We note that in the early 1980s, the advanced materials industry was in its infancy, as a number of products registered less than a few million dollars of sales globally. In fact, the entire industry could scarcely bring in $2 billion worth of sales world-wide, including purchases by research organizations as well as companies. In these years, advanced ceramics, polymers, composites, and alloys were the major players in the field. By 2030, even as these groups continue to diffuse into the economy, the other, more cutting-edge materials will have been commercialized to find their place in global markets. As is seen in Table 2.8, by this time, we project that the global (direct) market for advanced materials will reach $317 billion (with the indirect economic impact being three to four time this amount). For its part, the United States will capture $82 billion of this global market by 2030 (or over $410 billion if direct and indirect impacts are considered). By 2030, the United States will control 26% of the world market for advanced materials, down from an earlier 45%, as dictated by the convergence assumption (see Table 2.9).

Because of convergence, the European market ought to hold onto a progressively larger percentage of the global market. This is so because during this period the European Union will have been expanding its borders and integrating its members into a unified, coherent market economy with a size and reach on par with that of the United States. Then too, we assume that globalization has a "flat-

TABLE 2.8. "Potential" projected market for new materials, global—"convergence" model ($ billion)

	1980	1990	2000	2010	2020	2030
Bioengineered materials	—	—	0.5	1.2	5.6	10.0
Advanced ("super") alloys	0.4	3.8	15.4	25.4	30.2	35.0
Advanced ceramics	0.8	1.1	3.8	6.7	10.7	18.5
Advanced engineering polymers	0.3	6.0	10.6	15.3	22.5	31.4
Organic polymer electronics (OPEs)	—	—	0.1	0.4	5.9	25.4
Advanced electronic materials (other)	0.2	4.6	10.0	15.8	24.1	33.7
Advanced coatings	—	0.5	1.8	4.7	8.6	20.9
Nanopowders	—	0.6	2.5	10.6	23.5	55.2
Nanocarbon materials	—	—	0.4	4.8	12.6	27.4
Nanofibers	—	—	0.3	1.6	4.8	18.5
Thin films	—	0.2	0.5	3.6	8.5	12.5
Advanced composites	0.3	0.7	5.8	12.6	20.0	28.2
Total	2.0	17.5	51.7	102.7	177.0	316.7

Source: [17].

tening" effect across regional economies. The rise of multinational firms in conjunction with the evolution of the Internet and advanced communications and information systems opens distant markets to European products and processes. More than that, it accelerates the international transfer and application of advanced-material technologies to the extent that we must speak of "global" rather than "national" technological knowledge and skills. This will serve as a catalyst to

TABLE 2.9. "Potential" projected market for the new materials, United States—"convergence" model ($ billion)

	1980	1990	2000	2010	2020	2030
Percentage of global market (%)	35	40	45	40	30	26
Bioengineered materials	—	—	0.2	0.5	1.7	2.6
Advanced ("super") alloys	0.1	1.5	6.9	10.2	9.1	9.1
Advanced ceramics	0.3	0.4	1.7	2.7	3.2	4.8
Engineering polymers	0.1	2.4	4.8	6.1	6.8	8.2
Organic polymer electronics (OPEs)	—	—	—	0.2	1.8	6.6
Advanced electronic materials (other)	0.1	1.8	4.5	6.3	7.2	8.8
Advanced coatings	—	0.2	0.8	1.9	2.6	5.4
Nanopowders	—	0.2	1.1	4.2	7.1	14.4
Nanocarbon materials	—	—	0.9	1.9	3.8	7.1
Nanofibers	—	—	0.1	0.6	1.4	4.8
Thin films	—	0.1	0.2	1.4	2.6	3.3
Advanced composites	0.1	0.3	2.6	5.0	6.0	7.3
Total	0.7	6.9	23.8	41.0	53.3	82.4

the development of Europe's advanced-material capability so that it can compete on the world's stage as an equal with the United States. Our model assumes that the European Union will capture a growing proportion of these markets until 2020, after which, according to the law of convergence, competition from Asia will begin to take its toll on Europe. Accordingly, we should begin to see a contraction of the European Union's market share between 2020 and 2030. Nevertheless, over the entire half century between 1980 and 2030, we expect—assuming convergence—that the European Union will regularly approach the market share of the United States with respect to the pivotal advanced-material market. These assumptions are reflected in the following table, which shows that, for each decade beginning in 1980, the European Union market share and, consequently, the sales of its advanced-material industry (in terms of real dollars), approaches ever nearer to that of the United States.

These trends depend on the belief that the world is a fairer, more competitive place because of globalization. It certainly should be expected that if globalization is indeed the "flattener," as many today believe, we should at the least expect that two large, industrialized, and closely intertwined economies should, over time, be able to meet as equally effective players on the world economic stage. But is this

TABLE 2.10. "Potential" projected market for the new materials, Europe—"convergence" model ($ billion)

	1980	1990	2000	2010	2020	2030
Percentage of global market (%)	12	15	18	23	28	25
Bioengineered materials	—	—	0.1	0.3	1.6	2.5
Advanced ("super") alloys	—	0.6	2.8	5.8	8.5	8.8
Advanced ceramics	0.1	0.2	0.7	1.5	3.0	4.6
Engineering polymers	—	0.9	1.9	3.5	6.3	7.9
Organic polymer electronics (OPEs)	—	—	—	0.1	1.7	6.4
Advanced electronic materials (other)	—	0.7	1.8	3.6	6.7	8.4
Advanced coatings	—	0.1	0.3	1.1	2.4	5.2
Nanopowders	—	0.1	0.5	2.4	6.6	13.8
Nanocarbon materials	—	—	0.1	1.1	3.5	6.9
Nanofibers	—	—	0.1	0.4	1.3	4.6
Thin films	—	—	0.1	0.8	2.4	3.1
Advanced Composites		0.1	1.0	2.9	5.6	7.1
Totals	0.1	2.7	9.4	23.5	49.6	79.3

Note 1: The difference for each advanced material category and for the total between the "World" and the sum of United States and Europe represents the growing influence of Asia as an advanced-technology market.

Note 2: These market trends for the world, United States, and Europe refer to the markets for the materials only. If we also were to figure in the value of the final products that consume and incorporate these materials, then these numbers would have to be multiplied by a factor of three to four resulting in a *total* global market between $1 and $2 trillion by 2030.

indeed the case? Are the very reasonable numbers that we projected above in fact reality? Can we actually say that the technical and economic trends show the United States and European Union coming together as competitors? This question touches closely on the linkage between technology, productivity, and economic growth, and on the central role of the modern advanced-materials industry in creating national and regional competitive advantages at the end of the twentieth and into the twenty-first centuries. We begin to consider these matters in the following chapter.

REFERENCES

1. This table was constructed from a number of sources, including Spitz, P. H. (1989), *Petrochemicals: The Rise of an Industry,* New York, Wiley; Moskowitz, S. L. (2002). *Critical Advanced Materials Report: Final Draft* (Prepared for: Virginia Center for Innovative Technology, Herndon, VA); and Uldrich, J. (2006) *Investing in Nanotechnology,* Avon, MA: Platinum Press).
2. Shanley, A. (1997), "Biotech's New Mandate: More, Cheaper, and Faster—Chemical Engineering is Speeding Scale-up and Process Development," *Chemical Engineering,* Vol. 104, No. 1, pp. 1–28; Sokhansanj, S. and Wright, L. (2002), "Impact of Future Biorefineries on Feedstock Supply Systems: Equipment and Infrastructure," presentation at the 2002 ASAE Annual International Meeting/CIGR XVth World Congress, July 28–32, Chicago; National Research Council (1999), *Biobased Industrial Products: Priorities for Research and Commercialization,* National Academy Press, Washington, DC; Jones, D. (2002), "DOE Funds Six R&D Projects to Ramp Up Biomass Refining," *Inside Energy,* November 4, p. 11.; "NERL, DuPont Team Up for Development of Biorefinery" (2003), *Oxy-Fuel News,* Vol. 15, No. 41 (October 13), p. 1.; "Cargill Eyes New Bio-Based 'Platform' Chemicals" (2003), *Chemical Week,* Vol. 165, No. 44 (December 10), p. 12.; "A 'Watershed' for Bioprocessing" (2003), *Chemical Week,* Vol. (February 5), p. 19.
3. Bowman, R. (2000), "Superalloys: A Primer and History," Supplement, 9th International Symposium on Superalloys, NASA Lewis Research Center and The Minerals, Metals, and Materials Society; Lowe, T. (2002), "Nanometals Revolution is Crucial in Sustaining the Metals Industry," *Small Times,* February 6; Valenti, M. (2002), "Building It Better—Advanced Alloys and Plastics for Automotive," *Mechanical Engineering,* Vol. 124, No. 3, pp. 54–57.
4. Design Engineering (2003), "Nanoceramics on the Rise," Home Section, April 24; Rajan, M. (2001), "US Piezoelectric Market Continues to Boom," Business Communications Company, Press Release, March 28 (Norwalk, Connecticut); pp. 24–27.
5. Design Engineering (2003), "Nanoceramics on the Rise," Home Section, April 24; Moskowitz, S. L. (2002). *Critical Advanced Materials Report: Final Draft*, pp. 24–27 (Prepared for: Virginia Center for Innovative Technology, Herndon, VA).
6. Valenti, M. "Building It Better—Advanced Alloys and Plastics for Automotive," *Mechanical Engineering,* Vol. 124, No. 3, pp. 54–57; Tullo, A. H. (2003), "Engineering Polymers," *Chemical & Engineering News,* Vol. 81, No. 22, pp. 21–25.
7. Moore, S. K. (2003), "Just One Word—Plastics," Special R&D Report, IEEE Spectrum On-Line (www.spectrum.ieee.org), September 9; Klauk, H. (2000), "Molecular Electronics on the Cheap," *Physics World,* January, pp. 18–19; Leeuw, D. (1999), "Plastic

Electronics," *Physics World,* March, pp: 31–34; Goho, A. (2003), "Plastic Chips: New Materials Boost Organic Electronics," *Science News,* Vol. 164, No. 9, p. 133; Forrest, S., Burrows, P., and Thompson, M. (2000), "The Dawn of Organic Electronics," *IEEE Spectrum On-Line* (spectrum.ieee.org), Vol. 37, No. 8 (August); Teltech Resource Network Corporation (1999), "Organic Polymer Electronics: Phase 1," Teltech Publication, Minneapolis, MN; Boulton, C. (2003), "Researchers Developing Plastic Memory Technology," *Internetnews.com,* November 14.

8. Paint and Coatings Association Website (www.paint.org/ind_info/) (2001), Industry Information: 1996–2000; Moskowitz, S. L. (2002). *Critical Advanced Materials Report: Final Draft,* pp. 4–16 (Prepared for: Virginia Center for Innovative Technology, Herndon, VA).
9. EPRI Center for Materials Production (2000), *Industry Segment Profile: Composites* (Columbus, Ohio), pp. 5–32; Leaversuch, R. (2001), "Nanocomposites Broaden Roles in Automotive, Barrier Packaging," *PlasticsTechnology Online* (www.plasticstechnology.com), October.
10. Stuart, C. (2003), "Nanotubes are Bidding for Star Billing on Big Screens," *Small Times,* September 12; Brauer, S. (2002), "Online Exclusive: Emerging Opportunities for Carbon Nanotubes," *Ceramic Industry Online* (www.ceramicindustry.com), January 1; Holister, P., Harper, T. E., and Vas, C. R. (2003), *White Paper: Nanotubes,* CMO Cientifica (Las Rozas, Spain), pp. 1–13; Moskowitz, S. L. (2002). *Critical Advanced Materials Report: Final Draft,* pp. 16–24 (Prepared for: Virginia Center for Innovative Technology, Herndon, VA).
11. Pao, W. C. (2003) "Nanotech in Textiles to Hit $15 Billion," *The China Post,* March 27; Moskowitz, S. L. (2002). *Critical Advanced Materials Report: Final Draft,* pp. 27–30 (Prepared for: Virginia Center for Innovative Technology, Herndon, VA).
12. Moskowitz, S. L. (2002). *Critical Advanced Materials Report: Final Draft,* pp. 30–34 (Prepared for: Virginia Center for Innovative Technology, Herndon, VA).
13. Leaversuch, R. (2001), "Nanocomposites Broaden Roles in Automotive, Barrier Packaging," *PlasticsTechnology Online* (www.plasticstechnology.com), October; Scala, E. P. (1996), "A Brief History of Composites in the U.S.—The Dream and the Success," *JOM,* Vol. 48, No. 2, pp. 45–48; Kruschandl, N. (2006), "Composite Materials," www. solarnavigator.net/composites/composite_materials.htm; EPRI. Center for Materials Production (2000), *Industry Segment Profile: Composites* (EPRI: Columbus, OH).
14. Moskowitz, S. L. (2002). *Critical Advanced Materials Report: Final Draft* (Prepared for: Virginia Center for Innovative Technology, Herndon, VA).
15. These ideal life cycle patterns are estimated by adjusting the standard Gaussian life cycle model, well known to students of product innovation, with the particular characters of the various advanced material categories. These are determined from the same sources listed in this chapter (see References 2–14 above for sources used).
16. For detailed data on estimated Department of Defense (DOD) annual purchases of materials, components, and systems, refer to DOD's Defense Economic Impact Modeling System (DEIMS). For a more general discussion, see National Science Foundation (2005), *Advanced Materials: The Stuff Dreams are Made of,* Washington, DC, NSF, pp. 1–2.
17. Estimated and projected markets (in $ billions) were determined from a number of sources. In addition to the sources listed in References 2–14, a number of other sources were consulted. These include: Thayer, A. M. (2000). "Firms Find a New Field of Dreams: . . . Markets are Emerging for Nanomaterials," *Chemical and Engineering*

News, October 16.; Holister, P. (Ed.) (2003), *The Nanotechnology Opportunity Report,* 2nd edition, Executive Summary, CMP Cientifica, Lux Capital Group LLC; Graff, G. (2003), "The Nanomaterials Market is Starting to Climb the Growth Curve," *Purchasing Magazine,* August 28; Service, R. F. (2000), "New Age Semiconductors Pick up the Pace," *Science,* January 21, pp. 415–417; Bernard, A. (2003), "Handful of Industries at Leading Edge of Nano Commercialization," *Small Times,* July 29.

Chapter 3

The Great Divide: Advanced Materials, Productivity, and Economic Growth in the United States and Europe

A fundamental question in this story is how well technological innovation is being distributed between countries. We know that the United States dominated the first advanced-materials revolution in terms of where new technology originated. In this important sense, that great period of creativity between the late 1930s and the late 1950s is justifiably called the American technological revolution. But is this the case when we proceed ahead in time and examine the latest materials revolution from the 1960s through the early twenty-first century? After all, as already noted, it was in this period that globalization really accelerated its pace. This "taking off" point is no more in evidence than in the case of the European Union. In these decades, the European Union formed itself into a coherent body of states and moved progressively to the single market that it had been aiming at since the end of World War II. If the United States in its formative period during the first half of the nineteenth century could gain control of the industrial revolution through new technology, would we not then expect to see the Europeans regain the advantage in matters technological, especially given Europe's well-known scientific capability?

In fact, recent investigations by European organizations, public and private, point to the United States maintaining a healthy lead over Europe in technological development generally. From the 1970s to 2002,the United States exceeded the European Union in the sixteen major indicators of technological growth, including percentage of total R&D funded by industry, growth of the technological infrastructure, and other measures of technological primacy (see Figure 3.1).

In fact, the European Union's position in terms of high-technology trade has deteriorated. The European Union's trade deficit in high-technology products increased from EUR 9 billion to EUR 48 billion between 1995 and 2000. This gap continues to grow, even in the face of America's overall trade deficit.

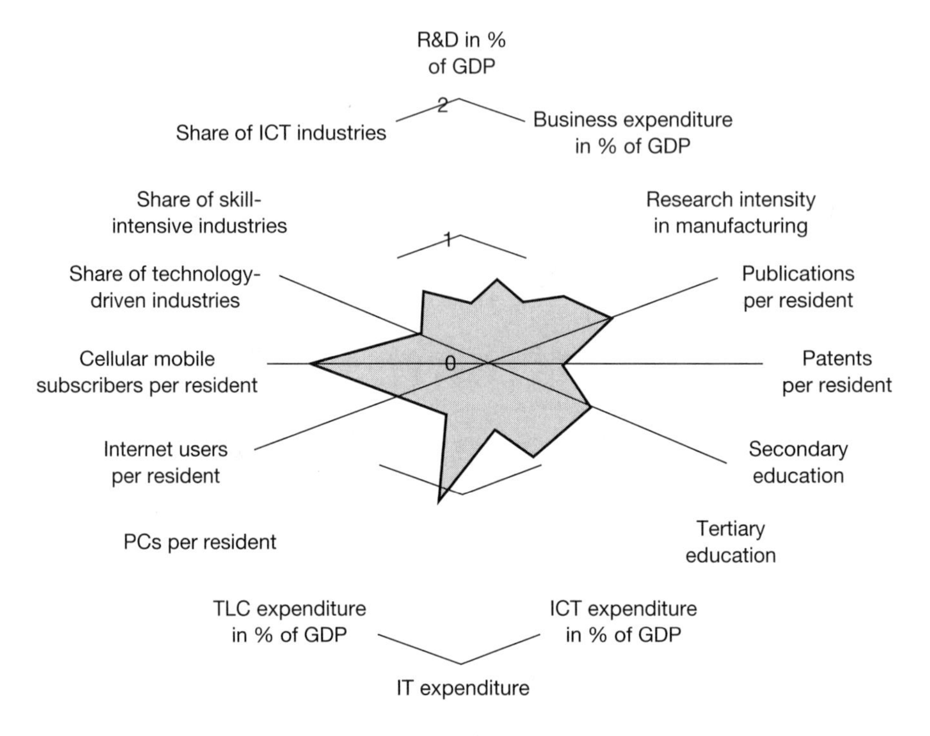

Figure 3.1. The United States leads according to most growth drivers. (Source: [1].)

Whereas these trends relate to technological achievement in general, they apply equally to the advanced-materials sector. The following table, which shows the year and national origins of the important advanced-material innovations from the 1950s to 2007, clearly tells us that the United States retains solid leadership in advanced-materials technology throughout the period.

We might go further and ask whether many of these innovations, though officially developed in the United States, are not the product of foreign firms that come to the United States in order to develop their basic patents. This technology transfer from Europe to the United States occurred in the past. For example, one of the most important petroleum refining technologies of the 1940s, fixed-bed catalytic cracking, was conceived and developed by the Frenchman, Eugene Houdry. In order to commercialize that process, he came to the United States looking for investment money and equipment to scale up his designs. Sun Oil Company (of Marcus Hook, Pennsylvania), owned and managed at the time by the Pew family, provided these and together Sun and Houdry successfully put their process online in a relatively short period of time. But this does not appear to be the case for the major innovations we are now considering. Advanced-materials technology of the late twentieth and early twenty-first clearly origninated and were developed by U.S. companies. In some cases, the firm (such as IBM, Union Carbide, Dow Chemical) accomplished this in house; in other cases, these larger firms worked patents originating in

Table 3.1. Major advanced-material landmarks: 1950s–2002

Innovation	Year	Country
Germanium-based semiconductors	1950s	USA
Diffusion furnace for doping silicon wafers	1950s	USA
Landmark paper on possibility of fabricating materials atom by atom	1959	USA
Large single crystals of silicon	1960	USA
Magnetic cards (for computers)	1960	USA
Directionally solidified (DS) "super alloys"	1960	USA
Nickel–titanium alloy shape memory	1962	USA
Semiconducting materials: slicing and doping of silicon crystals	1962	USA
Carbon fiber	1964	UK
Thin-film resistor materials	1965	USA
Multilayer metallization	1965	USA
Advanced composites: high-modulus whiskers and filaments	1960s	USA
Optical fibers	1970s	USA
Single-crystal (SC) "super alloys"	1970s	USA
Microalloyed steel	1970s	USA
Amorphous metal alloys	1970s	USA
Kevlar plastic	1973	USA
Metal matrix composites	1974	USA
First molecular electronic device patent	1974	USA
Solid-source molecular-beam epitaxy (MBE)	1975	USA
Polymer metal composites: boron Filaments	1975	USA
Metal and polymer–metal composites: silicon carbide fibers	1976	USA
Polymer–metal composites: graphite-reinforced composites	1976	USA
Electrically conducting organic polymers	1978	USA
New generation of deep UV-photoresist materials (for advanced semiconductor lithographic techniques via "chemical amplification")	1980	USA
Rare-earth metal application to semiconductors	1980s	USA
Lanthanum–barium–copper oxide material discovered to be "superconducting"	1980s	Switzerland
Scanning tunnel microscope: atomic- and molecular-scale imaging	1982	USA
Gas-source molecular-beam epitaxy (MBE) for thin films	1985	USA
"Buckeyball" fullerenes discovered	1985	USA
New generation of optical polymers for flat-panel, liquid-crystal displays	1986	USA
New generation of piezoelectric crystals	1987	USA
New generation of electromagnetic materials for MRI technology	1987	USA
New generation of advanced stainless steels	1980s	USA
Landmark technique: scanning tunneling microscope precisely positions 35 xenon atoms to spell "IBM"	1989	USA
Synthetic skin	1990	USA
Carbon nanotubes discovered	1991	Japan
Vacuum arc–vacuum reduction stainless steel technology	1996	USA
High-purity single-walled nanotubes (via laser vaporization)	1996	USA

(*continued*)

Table 3.1. *Continued*

Innovation	Year	Country
First commercial biorefinery	1999	USA
Thin-strip casting for stainless steel	2000	USA
Superconducting "Buckminsterfullerene" crystals	2001	USA
Ultrathin-layer high-dielectric insulating materials	2002	USA
"Super" computers	2007	USA

Source: [2].

other U.S.-based companies; and in still others, an entrepreneur licensed the central patents from a U.S. government laboratory or, more often, a U.S. university as the anchor of his start-up company. Table 3.2 displays those firms, together with their national origins, that are considered the most important players globally in the advanced-material field.

We see that the great preponderance of these companies—over 80%—are based in the United States. These companies—small firms for the most part—arose in the 1980s and 1990s and constitute the cutting edge of the U.S. technological base. Their research and products have been diffusing throughout the U.S. economy over the last two decades. Even in this era of globalization, even with the spread of multinational companies worldwide, and even with U.S. and foreign firms collaborating together on R&D projects as never before, we still have not seen the "spreading out" of technological accomplishment. Neither the expansion and greater integration of Europe and its markets nor the rise of Asia as a potential economic power appears to have shifted or "flattened out" important technological activity.

But, after all, to know whether or not globalization spreads economic benefits more widely, is it enough to examine these trends in the source of technology? That is, can we conclude that a region with technological advantages, such as the United States appears to be, must also have a concentrated economic advantage? In other words, can globalization disperse economic activity across nations and regions even as technological prowess remains localized? Certainly, we might expect that, with the global integration of markets and the liberalization of trade internationally, economic activity and wealth creation would spread across countries and into different regions, even if they could not match the United States in the most advanced technological development. The case of the European Union, to take a most important example, presents us with a number of compelling indicators that would lead us to expect economic growth in Europe commensurate with the United States. According to Euobserver data, at purchasing power parity, the current (2008) 27 members of the European Union have a combined GDP slightly larger than that of the United States. The European Union is also around the same size as the United States in terms of trading power, with both the European Union and United States accounting for a similar share of global imports and exports. The European Union's market is immense. It has a combined population of 492 million, or 7.4% of the world's population. It ranks third in population behind China and India. Further, 13 of the 27

Table 3.2. The major advanced-technology firms: United States and global (2007)

Company	Country	Technology
Applied Nanoworks	USA	Nanocrystals
Arrow Research Corp.	USA	Nanocomposites
Carbon Nanotechnologies	USA	Nanotubes
Capsulation Nanoscience AG	EU (Germany)	Nanocapsules
Cargill Dow	USA	Biosynthetics
Crystal Plex Corp.	USA	Nanocrystal beads/nanocrystals
Eikos, Inc.	USA	Nanotubes
Elan Corp., PLC	EU (Ireland)	Advanced crystals
Evident Technologies, Inc.	USA	Nanocrystals/quantum dots
Flanel Technologies, Inc.	EU (France)	Nanoparticles
Hybrid Plastics, Inc.	USA	Advanced plastics
InMat Inc.	USA	Nanocomposite coatings
Isonics corp.	USA	Silicon nanomaterial
Lucent Technologies	USA	Organic electronic materials
Lumera Corp.	USA	Advanced polymers, electrooptical polymers
Luna Nanoworks	USA	Fullerenes
Monsanto	USA	Genetically modified organisms
Nano C	USA	Nanotubes
Nano-Tex	USA	Advanced fibers
Nano BioMagnetics, Inc.	USA	Advanced magnetic materials
Nanocor, Inc.	USA	Nanopowders/polymers
Nanodynamics, Inc.	USA	Advanced ceramics, nanotubes
Nanofilm, Ltd.	USA	Ultrathin Coatings
Nanogate Technologies GmbH	EU (Germany)	Multifunctional materials, nanopowders
Nanomat	USA	Nanopowders
NanoProducts Corp.	USA	Catalysts/nanopowders
Nanoscale Materials	USA	Metal oxides/nanopowders
Nanostellar	USA	Composites/catalysts
Nanotechnologies, Inc.	USA	Nanopowders
Optiva, Inc.	USA	Thin-crystal film
Oxonica	EU (UK)	Nanopowders
Polycore Corp.	USA	Nanocomposites, nanoclay
Qainetiq Nanomaterials, Ltd.	EU (UK)	Nanomaterials
Quantum Dot Corp.	USA	Quantum dots
Quantum Sphere	USA	Nanomaterials
Reactive Nanotechnologies	USA	Advanced binding materials, nanofoil
Showa Denko K.K.	Japan	Carbon nanotubes
Southwest Nanotechnologies, Inc.	USA	Carbon nanotubes
SuNyx	EU (Germany)	Nanomaterials
Starpharma	Australia	Dendrimers
Union Carbide	USA	Diverse materials

Source: [3]

members of the European Union have adopted the Euro, widely considered to be the only global currency that can (and indeed has been doing so since 2002) challenge the dominance of the U.S. dollar. In addition, an increasing number of multinationals from the European Union have joined their U.S. and Japanese competitors in the fortune Global 500 list of the world's biggest companies. The European Union also has a high Human Development Index, which exceeds that of the United States.

What we are asking here essentially concerns the relationship, if any, between technological change in general, and advanced-materials creation in particular, and the level of a country's economic performance in the late-twentieth and twenty-first centuries. If technological change closely shadows economic performance, and if we find that the European Union economy has been converging toward, and even exceeding, U.S. economic performance in recent years, then, given the fact that advanced technology continues to emanate within the United States, we can reasonably conclude that (1) globalization has been dispersing U.S. technical output to other industrial nations (i.e., Europe), and (2) these nations effectively translate this foreign (i.e. U.S.-created) technology into domestic economic growth. The following sections take up these issues, with the discussions centering on the relationship of advanced-materials technology to economic performance since the 1970s.

TECHNOLOGY AND ECONOMIC PERFORMANCE I—IT, ENERGY, AND BIOMEDICAL

There are essentially three ways in which a country can achieve robust economic growth: Citizens can work more, they can work more efficiently, or they can do both. All three of these ways increase annual earnings. In the case of working more efficiently, workers effectively utilize their work input. Increased productivity gains are immensely important in that they help to reduce inflationary pressures (as does monetary tightening and rising interest rates). Although the links between welfare, economic output, and productivity are complex in practice, as well as in theory, it is generally true that if productivity increases, all else being equal, aggregate economic welfare increases. Indeed, productivity trends are so vital to a nation that they are used to forecast potential economic activity. Long-term productivity growth is commonly viewed as the speed limit for sustainable economic growth. It is for this reason that the World Economic Forum concludes that "Competitiveness finds its ultimate expression in the prosperity that countries can sustain over time. Prosperity is sustainable if it is based on the productivity companies can reach given the conditions they face in an economy" [4].

But where does such productivity growth come from? Economists such as Alan Greenspan, Paul Krugman, and others clearly point to technological innovation as the critical factor in the rising tide of labor productivity in recent decades, especially in the United States.

The case of information technology is an important example of how innovation is adapted to business to accelerate productivity. In the 1980s and 1990s in particular, these innovations entered into and transformed office and factory information-

and data-processing systems. Memory and applications of computer technology expanded and new hardware and software products eased the adaptation of advanced systems to the workplace. These advances, in turn, led to enormous advantages in workplace productivity, especially in such areas as data processing, inventory control, just-in-time (JIT) delivery, and the like. The World Economic Forum (WEF) points to just this type of innovation-adaptation mechanism as the main driver of modern productivity and economic competitiveness:

> . . . [T]echnological differences have been shown to explain much of the variation in productivity between countries. . . . The relative importance of technology adoption for national competitiveness has been increasing in recent years. . . . The strong productivity growth recorded in the United States over the past decade has been linked to the high adoption of information technologies [by, for example] the retail and wholesale sectors [which use IT extensively]. [5]

It is estimated that by the turn of the twenty-first century, nearly three-quarters of economic growth depended on advancing technology, a trend that has been progressively moving upward since the 1960s. It is for this reason that the WEF finds that there is a strong "correlation between competitiveness rankings and a measure of new technology usage in a large number of countries [thus] underscoring the central importance of [advanced technology] for productivity" [6].*

From a close reading of business, economic, and government analyses and trends, we can estimate the percentage of economic growth attributable to advancing technology from 1970 to 2000. We can also project out this percentage from 2000 through 2030 (Table 3.3).

The three leading sectors that drive, or will drive, productivity growth are information and communications technology (ICT), energy, and biomedical and healthcare. ICT today is probably the most important reason for productivity growth in the industrialized world since the 1980s. Jorgenson, Ho, and Stiroh (2000) argued that accelerated innovation in ICT has altered the speed of productivity growth in the United States, Europe, and other developed nations [22]. Accordingly, If technical advances slow down, so will productivity growth globally. We know, for instance, through Moore's Law, that technical improvements have doubled capacities of computer memory and disk storage and halved microprocessor feature sizes roughly every two or three years since the 1980s. They have also reduced the costs of manufacture and distribution, thus allowing price reductions. These technical advances have had an enormous impact on productivity. In a paper written in 2000, Drs. Brynjolfsson and Hitt of, respectively, MIT and the University of Pennsylvania, explain that the strong linkage between these very rapid advances in capacities, memory, miniaturization, and productivity derives from the fact that "Information technology, defined as computers as well as related digital communication technology, has the broad power to reduce the costs of coordination, communications, and

*The WEF's annual "Global Competitiveness Rankings," which ranks countries according to their overall ability to compete in the global economy, heavily considers such indices as "technological readiness," "innovation," and "higher education and training" in determining their rankings.

Table 3.3. Percentage of economic growth attributable to advanced technology

Year	Technology in general (%)
1970	40
1980	55
1990	65
2000	73
2010	78
2020	85
2030	90

Source: [7].

information processing" [8]. The authors estimate that rates of return on investments in ICT has been approaching 50% annually on a company basis.

This productivity growth through new ICT technology does not occur immediately upon the availability of the new products and systems. As Dr. Paul David of Stanford explains, the reason we did not see impressive productivity gains in the US until the 1990s was because society has to wait for diffusion and adaptation of new technology to take hold. Successful adaptation often requires organizational changes to be made, new skills developed, and experimentation on possible optimal application carried out [23].

Although ICT continues to dominate productivity growth within the industrialized nations, energy and biotechnology are advancing as important drivers as well. Energy technology, in particular, is of especial importance, particularly since the 1970s. Productivity in this case is measured in terms of units of energy used to make a unit of product or service. The importance of energy efficiency to an economy is reflected in annual energy intensity trends within the United States from the 1950s to the turn of the 21st century. The improvements since 1970 of overall energy efficiency of the U.S. economy resulted from a complex and changing mix of increases in the efficiency of energy transport; oil refining; electricity generation, transmission, and distribution; and household and commercial appliances. Table 3.4 shows trends in the annual rate of energy intensity in the U.S. economy. As is seen, there have been significant declines in energy intensity every decade, starting in the 1970s.

Biomedical technology is the latest sector to influence productivity growth. This is so because biomedical advances increase the quality of the performance, and extend the useful working life, of a society's workforce.

Table 3.5 estimates the percentage of economic growth attributable to each of these sectors from 1970 to 2000, and projects these percentages from 2000 to 2030. The table shows the rapid rate at which ICT technology entered into U.S. manufacturing and services between 1980 and 2000, and its continued role as a driver of productivity after 2010. The trend indicates the rise of energy and biotechnology as

Table 3.4. Annual rate of energy intensity: U.S. economy

	Annual energy intensity
1955–1970	Constant
1970–1980	–1.7%
1980–1985	–3.5%
1985–1995	–1.0%
1995–2000	–2.7%

Source: [9].

forces for productivity growth through 2030. We also note the steady decline over these six decades in the contribution made by other (i.e., non-technically related) factors.

TECHNOLOGY AND ECONOMIC PERFORMANCE. II—THE ROLE OF ADVANCED MATERIALS [11]

If, as we contend, economic growth and technological advance go hand in hand, especially in the decades leading to the twenty-first century, what role do advanced materials play in economic activity? Are they but the handmaidens of IT, energy, and biotechnology, one of a number of technical inputs that these three sectors depend upon for their continued progress, or is the advanced-materials industry absolutely central to the status they enjoy as leading drivers of economic growth? To address these questions, we examine in more detail the function and importance of advanced materials within these critical sectors.

Within the ICT industry, it is certainly true that improvements in functionality of semiconductor chips, computers, and communications systems require skillful organization of the transistors on the chips, innovative hardware designs, and creative software programs. But these would not be possible without the materials that form the technical superstructure through which new arrangements of components prove effective and upon which advanced software programs can operate. Most fundamentally, materials innovation must continue to occur in order to handle increased

Table 3.5. Impact of advanced technology on U.S. economic growth: distribution by sector—ICT, energy, and biotechnology (% contribution)

	1970	1980	1990	2000	2010	2020	2030
ICT	15	20	35	55	60	65	68
Energy	3	5	8	10	12	15	17
Biotechnology	1	2	5	5	5	10	12
Other	81	73	52	30	23	10	3

Source: [10].

power densities and heat dissipation as chips continue to evolve. Most often, semiconductors and computers compete on the basis of such technical characteristics as size and speed of transistors and internal data bandwidths. The compiler, which converts high-level programs into code that actually runs on the microprocessor and uses its transistors, is often the main source of processing power. Between 1955 and 1990, improvements and innovations in semiconductor technology increased the performance and reduced the cost of electronic devices by a factor of one million. These technical advances depended on the discovery and application of new material systems.

Within the IT sector, materials have played, and will continue to play, a dominant place in technological advance. These materials continually drive forward semiconductor and photoelectronic technology. In their study conducted in the late 1990s, Kim and Kogut assert the close relationship between materials innovation and semiconductor technology. In tracing the evolutionary trajectory of semiconductor product development, they find that an important ". . . way to distinguish the evolutionary [course of semiconductors] is by tracing the evolution of . . . [advanced] materials as 'co-evolving' with [semiconductor] applications." [24] The semiconductor crystals themselves posed a challenge to the materials industry. New grades and compositions of silicon crystal materials continue to develop, but materials technology drives the entire semiconductor process. The technique of patterning polished wafers with an integrated circuit absolutely demands the use of photoresist materials in thin-film configurations. These materials, which are advanced polymers modified by radiation that contain photoactive compounds, form thin coatings on the semiconductor wafer and control the photolithographic process. The technical challenge here, as the chip becomes smaller, is to find novel photoresist materials to accommodate shorter wavelengths of light, required to accurately define the smaller features of the microchip.

Beyond the crystal and patterning issues are the important problems involved in establishing electrical connections on the chip and packaging the semiconductor device. Interconnection technology is a critical and difficult part of the semiconductor production process. It is through various conductors and connectors that the separate microcomponents are linked into an integrated system. The capability of new types of connecting materials, in the form of metallized film, considerably advances connecting techniques. This is important because the capacitance involved in the traditional "component-followed-by-component" configuration retards the flow of electrons and, therefore, of information within the chip. New material design now allows the integrating of several chips into a single "multichip module." In this approach, the chips are joined together on a shared substrate by the connecting materials. Multichip modules are made up of as many as five microchips bonded to a silicon or ceramic substrate on which resistors and capacitors have been constructed with thin films. Typical material and material application techniques used in multichip modules involve gold-paste conductors applied in an additive process (like silk-screen printing), the use of innovative glazes to insulate the gold-paste conductors from subsequent film layers, and the application of a series of thin films made with tantalum nitride, titanium, palladium, and plated gold.

Since the 1970s, opto- and photoelectronic technology has formed an integral part of ICT systems. Technological interdependence between these two fields is so strong that we find the photonics field often driving advances in ICT systems. Such devices as lasers, light-emitting diodes (LEDs), photodetecting diodes, optical switches, optical amplifiers, optical modulators, and optical fibers propel information movement and control, business and management system development, and computer hard- and software innovation. New materials development fuels photonics innovation. Photonics technology depends vitally on the development and novel application of high-grade semiconductor materials including gallium arsenide, aluminum gallium arsenide, indium phosphide, and aluminum indium arsenide. The process by which these compounds are made is crucial because fabricating a single crystal from these combinations of elements proves significantly more difficult than forming a single crystal of electronic grade silicon. Defect rates due to thermal stresses in the furnace prove a significant bottleneck since defects reduce the effectiveness of the photonic components. Accordingly, improving the technology of the furnaces in which these compounds are made is a paramount issue in the field.

As with silicon-based technology, photoelectronic development hinges not only on the central semiconductor crystals themselves, but on other, equally important components. For example, photonic materials include semiconductors for light emission or detection; elemental dopants that serve as photonic performance-control agents; metal- or diamond-film heat sinks; metallized films for contacts, physical barriers, and bonding; and silica glass, ceramics, and rare earths for optical fibers. Even more advanced material technology infuses two of the most important recent development in photonics: ultrathin "epitaxial" layer systems and optical switching techniques, both of which, in turn, cannot progress without development of ever new and more powerful materials and their process technologies.

Ultrathin layer materials technology is the sine qua non of efficient emission or detection of photons in optoelectronic devices. In a technique known as "band-gap" engineering, these thin semiconductor layers (or epitaxial layers) are grown on top of thicker (or bulk) semiconductor crystals; both layers are composed of the same or similar materials. Thus, for example, a thin layer of gallium aluminum arsenide might be formed above a gallium arsenide crystal. This sandwiching and repeating of very thin layers of a semiconductor between layers of a bulk crystal creates totally unique and vital materials that allow the modification of the band gap of the sandwiched layers critical in the functioning of advanced photonic and optoelectronic devices. As with the growth of crystals themselves, the economic formation of such thin layers requires innovative process technology. In particular, molecular-beam epitaxy (MBE) is the most advanced and precise method of growing epitaxial layers on a semiconductor substrate and is a large factor in the evolution of optoelectronic and photonic integrated circuits from the R&D department to the marketplace. In this process, a stream or beam of atoms or molecules moving in a vacuum strikes a heated crystal surface, forming a layer that has the same crystal structure as the planar substrate. Further advances in this field must address an economic and reliable method of growing epitaxial material on nonplanar structures, for example, around ridges or in "tubes" and "channels" that are etched into the surface of semiconducting devices.

Although these innovations, driven by new materials development, accelerate transmission of more information at higher speeds, there is still the need to convert this data from the optical to the electronic context. Development here focuses on the switching of information and data streams with ever-higher speed efficiency. But the reality is that the electronic form of switching has reached a barrier to increased speeds, and photonic methods offer possible routes to overcome this limitation. One recent device for photonic switching is the quantum-well self-electrooptic-effect device (SEED), consisting of many thin layers of two different semiconductor materials.

Recent technical developments in both energy and biotechnology make use of these recent developments in electronic and photonic innovations. These technological linkages extend out in many directions. The importance of rapid data and information transmission within the international biomedical community; the design of advanced sensor systems to increase fuel efficiency of automobiles and to detect leaks in petroleum, gas, and chemical piping networks; the use of remote work stations, wireless instruments, and local area network systems within large hospital and medical research complexes illustrate these electronic and information applications. But, even without reference to electronics, advanced materials play a growing role in both the energy and biomedical fields.

The relationship between materials and energy is complex and pervasive. At every stage of energy production, distribution, conversion, and utilization, materials play key roles, and special material properties are often needed to achieve higher efficiencies in energy usage.

Energy materials can be either passive or active. Passive materials do not take part in the actual energy conversion processes; rather, they facilitate production, transportation, storage, and distribution of petroleum, natural gas, fuel, and other forms of energy. Passive materials are used in containers, tools, or structures such as reactor vessels, pipelines, turbine blades, oil drills (diamond), and so forth. Active materials take part directly in energy conversion. They are used in solar cells, batteries, catalysts, superconducting magnets, biomass conversion, hydrogen fuel, and fast breeder reactors. In the construction industry, new materials and electronic and related technology have significantly increased energy use efficiency over the last three decades. In power generation and manufacturing, new materials come into play in numerous applications. To take but one example, new, more efficient microturbines depend on advanced materials for the combustion systems to reduce emissions by 50% while increasing efficiency by up to 10%. New types of catalysts and fuels drive energy efficiency in the transportation industry.

Within the biomedical field, new materials have become the essential components for advances in implants, grafts, drug-release systems, and biosensors. Novel types of polymers, metals, ceramics, and composites continue to push the envelope in implant technology, and are used in heart valves, hip replacements, breast implants, and so forth. As they are foreign objects functioning within the body, such materials must satisfy numerous requirements simultaneously, including hardness, tensile strength, fatigue strength (in response to cyclic loads or strains), resistance to abrasion and wear, long-term dimensional stability, and permeability to gases, liquids, and solids. There must also be compatibility between body and material to prevent such dangerous conditions as thrombosis (i.e., blood coagulation and adhesion of blood platelets to biomaterial surfaces). For orthopedic devices (such as hip-joint replacements), the earlier application of traditional metals and polymers is rapidly giving way to the new generation of materials. For instance, stainless steel was first employed but caused problems due to corrosion. Newer materials then came into play, including titanium and cobalt–chromium–molybdenum alloys and carbon-fiber reinforced polymer composites. Newer types of polymer–matrix composite materials, based on the polysulfone and polyetherketone polymers, now exist that allow implants to better distribute stresses.

Grafting and biosensor technology also proceeds only as suitable materials become available. Vascular grafts in the small-diameter region (such as in the legs), for example, require materials that minimize the dangers of thrombotic occlusion. Similarly, the barrier to further progress in implantable miniature biosensors technology, designed to measure a wide range of blood conditions continuously and provides critical contributions to medical diagnosing and monitoring, depends upon finding appropriate biomaterials. Advances in the manipulation of molecular architectures at the surface of materials by using chemisorbed or physisorbed monolayer films, combined with the development of nanoscale probes that permit examination at the molecular and submolecular level, are the collective key to the new wave of implantable biosensors.

This overview suggests that advanced materials technology infiltrates those three sectors most pivotal to economic growth. From interviews conducted with industry and technical personnel and from a variety of secondary sources, we may estimate and project the percentage of growth of each of the three sectors attributable to advanced materials from 1970 through 2030. These are shown in Table 3.6. As is seen, advanced materials have an earlier and greater impact on the ICT sector. By 2020, more than 75% of ICT's growth could be attributable to advanced materials, and by 2030 we project that this figure will reach 85%.

The growing importance of new materials for the energy and biotechnology sectors continues as well throughout the period.

From these trends, we estimate the percentage of economic growth attributable to advanced materials (with more weight placed on ICT given its greater impact on the economy). From these trends, we predict that the advanced materials industry

Table 3.6. Impact of advanced material technology on the ICT, energy, and biotechnology sectors (% contribution)

	1970	1980	1990	2000	2010	2020	2030
ICT	15	25	40	55	65	75	85
Energy	10	15	30	45	55	65	70
Biotechnology	5	10	20	30	45	55	65

Source: [12].

will account for more than half of economic growth somewhere between 2000 and 2010. By the end of the period in 2030, over three-quarters of all economic growth will be attributable to the development and application of advanced materials (Table 3.7).

LOCALIZATION, GLOBALIZATION, AND THE COMPETITIVENESS FACTOR: EUROPEAN UNION VERSUS UNITED STATES

Comparing Europe with the United States, we see that globalization has not yet leveled technological accomplishment; technical creation appears to remain discretely concentrated within the United States and, moreover, very much rooted in specific sectors of the economy. We also observe the central importance of advanced materials within the realm of modern technology. We have measured this significance through the increasing contributions of new materials to technological progress in general. Moreover, we have seen the strong nexus existing between an advancing materials technology and a nation's productivity performance and, in turn its economic growth.

At this point, we must tread somewhat carefully. Can we assume that that a technologically creative nation or region must also enjoy greater economic activity? In other words, if globalization does not disperse technological creativity, can it nevertheless spread the fruits of a creative nation and, thus, economic growth, to other parts of the world? If this were true, then the economic leadership of the original creators of new growth-inducing technology, especially industrialized nations, should narrow significantly over time. This would be especially true since the 1980s, when globalization created numerous opportunities for international technology transfer such as international joint ventures, licensing, cooperative R&D agreements, and foreign direct investments.

In fact, in comparing economic growth within the United States and the Euro-

Table 3.7. Percentage of economic growth attributable to advanced-material technology

Year	Technology in general (%)	Advanced-material technology (%)
1970	40	12
1980	55	20
1990	65	34
2000	73	48
2010	78	60
2020	85	70
2030	90	78

Source: [13].

pean Union, we find that the creators of new technology are the most economically active as well. Figure 3.2 shows production trends for the United States and the European Union up to 2000. Clearly, the United States has pushed ahead of Europe, especially during the 1990s. Productivity trends tell the same story. In the period 1970–1994, U.S. productivity (in terms of output per hour) was 1.4%. But between 1995–2000, productivity surged to 2.4%. This rapid growth in productivity has led some economists to label the period of the 1990s as ushering in the "New Economy." In terms of GDP per worker and GDP per hour, the United States, by 2000, exceeded European productivity by 20%. Starting in 1990, growth of real GDP in the United States was greater than that of both the European Union and Japan by 25% to 30% [14].

These productivity differences translate into significantly different growth patterns. Within the European Union, slack productivity relative to the United States prevents European resources from being generated and utilized as effectively as they should. Thus, critical resources are wasted that could be applied to augmenting opportunities for improving the material qualities of life and dampens prospects for robust economic growth. In contrast, the U.S. productivity situation increases private purchasing power and, through an expanding tax base, the prospects of heightened government services for the common welfare. It also reduces inflationary pressures since, with this expansion in productivity, the economy could (theoretically) grow 50% without fear of widespread price increases. All this leads to more

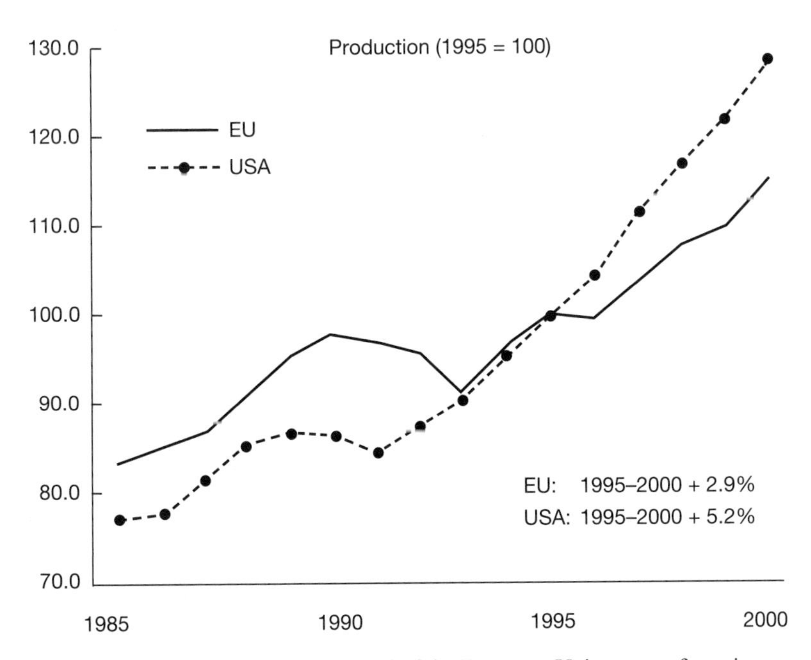

Figure 3.2. The United States forges ahead of the European Union: manufacturing—production (source: [18]).

economic growth. The American economy has been growing faster than the economies of many European countries and of the EU as a whole. From 1980 to 2005, the economic growth rate for the European Union averaged 2.1%, with a number of major economies such as Germany doing considerably worse on average, in comparison with an average annual rate of 3.0% for the United States.

Although the European Union has managed to avoid recession (at least through 2007), the economies of several of its major members, including Germany, France, and Italy, have stagnated or been lackluster. In fact, through the 1990s, the United States has led Europe (and Japan) in the major indicators of economic progress to the point that the U.S. economy has been the main engine of world growth. It is indeed telling that Swedish researchers from their own studies admit that, despite " . . . the elimination of trade barriers and the close economic integration resulting from the Single Market . . ." the European Union continued to lag behind, and significantly so, the United States as an economic performer. The authors go as far as to say that as "Productivity growth was lagging far behind . . . the United States, . . . the process of *convergence* in productivity, a much talked-about process since the 1970s, has once again become a process of *divergence*." [emphasis as in original] [15]. Indeed, the Swedes go as far as to conclude that "Perhaps Europe has something to learn from the United States when it comes to creating favorable conditions for an efficient market economy" [16]. A noted MIT economist recently considered European versus U.S. productivity trends: ". . . it looks like Europe is stuck fairly far behind the United States and the question is 'why [hasn't] the EU [caught] up?'" [17].

A number of factors account for this global discrepancy in productivity and economic growth. These include currency exchange rate movements as well as the fact that the European Union is constrained by inflexible labor markets, rising energy prices, and increased competition from Asia. Cultural differences regarding leisure time as an influence on productivity differences certainly receives considerable media attention. The significantly greater number of hours dedicated to work in the United States appears to be one component in America's *production* gains.

But these factors must not be overemphasized, for even with these variables removed from the equation, the divergence in productivity continues unabated, as is seen in Figure 3.3. These trends are supported as well by a plethora of qualitative and documentary evidence. The annual European Competitiveness Reports, for example, consistently underscore the widening productivity gap between the United States and Europe, especially in the 1990s and into the first decade of the twenty-first century [20].

As we have seen, in the late twentiety and twenty-first centuries, technology drives productivity and economic growth. In its business competitive Index rankings of the various countries of the world, the World Economic Forum considers factors that touch on innovation and companies' skills in generating, commercializing, and adapting new productivity-enhancing technology. These rankings consider the ability of businesses within a country to apply new technology toward economic growth, whether that technology is created at home or transferred from abroad. Based on these criteria, the World Economic Forum ranks the United

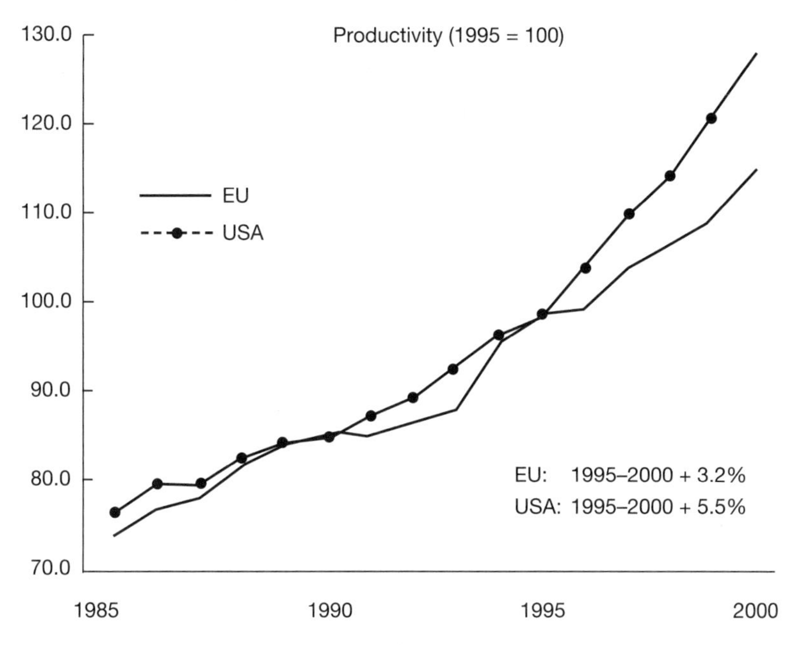

Figure 3.3. The United States forges ahead of the European Union: manufacturing—productivity (source: [19]).

States ahead of all the European Union countries. The rankings certainly show differences in the degree of competitiveness of the various European Union members. Switzerland, Denmark, and Sweden (not shown) are strong competitors occupying positions two, three, and four, respectively. But the major western European economies are significantly below the United States as global competitors. The overall picture that emerges is one of a European Union whose economy has been falling behind a growing, technologically vibrant United States.

We see, then, that globalization does not lead to convergence, even in the case of two large, industrialized, closely intertwined economies. We also understand that advanced-materials technology—more and more the hub of productivity growth—plays an increasingly key role in determining the economic growth potential and, therefore, competitiveness of a country or region. Based on a comparison of productivity trends and economic growth patterns between the United States and the European Union, and supported by international competitiveness rankings, Europe does not seem able to compete with the United States in creating new productivity-enhancing technology (in the form of advanced materials) and effectively harnessing these to the engines of economic growth. Indeed, even if the forces of globalization were to deliver new U.S. technology to the door of Europe, we need to consider that Europe is not as adept as the United States in identifying potential markets and guiding them as efficiently and effectively to a functioning reality.

If divergence, rather than convergence, is the proper way to understand the

Table 3.8. Global competitive index rankings: United States and European Union, the major western economies, 2007–2008

	2007–2008
United States	1
EU:	
Germany	5
UK	9
Austria	15
France	18
Belgium	20
Ireland	22
Spain	29
Portugal	40
Italy	46
Greece	49

Source: [21].

forces of globalization, then we need to question the model used to project the European market for advanced materials technology developed in Chapter 2. Europe, then, is not likely to realize this *potential* in large part because the European economy is not as successful as the United States in creating and integrating new products and processes to push forward productivity and, therefore, economic expansion and, ultimately, global competitiveness.

The chapters to follow focus on the forces behind this observed pattern of divergence. Given the decisive role played by technology in general, and advanced materials in particular, we turn to the different ways in which the United States and Europe manage the creation, development, and adoption of advanced materials innovation. It is vital to first understand the various risks that must be faced and dealt with. We can then proceed to ask how and in what ways the United States has been able to minimize these risks in order to proceed with successfully creating, diffusing, and adapting the most advanced and pivotal material technologies. We can then better appreciate how and in what ways Europe has failed to do so.

REFERENCES

1. Aiginger, K. and Wieser, R. (2002), *Factors Explaining Differences in Productivity and Competitiveness,* Austrian Institute of Economic Research (WIFO). Prepared for presentation at the Competitiveness Conference, January 30, 2002, Brussels, Belgium, p. 13.
2. See Chapter 1, Reference 3.
3. *NanoInvestor News* (www.nanoinvestornews.com) (2001–2005), Company Archives.
4. World Economic Forum (2006), *Global Competitiveness Report,* Geneva, Switzerland, p. xxiv.

5. World Economic Forum (2006), *Global Competitiveness Report,* Geneva, Switzerland, p. 10.
6. World Economic Forum (2006), *Global Competitiveness Report,* Geneva, Switzerland, p. 10.
7. Table is constructed from a number of sources, including: Holdren, J. P. (2001), "Energy Efficiency and Renewable Energy in the US Energy Future," Testimony for the Committee on Science, U.S. House of Representatives, February 28 (Washington, DC); Basu, S., Fernald, J. G., and Shapiro, M. D. (2001), *Productivity Growth in the 1990s: Technology, Utilization, or Adjustment,* Working Paper 8359, National Bureau of Economic Research (Cambridge, MA), pp. 1–6, 23–39; Tuomi, I. (2001), "Economic Productivity in the Knowledge Society: A Critical Review of Productivity Theory and the Impacts of ICT," firstmonday.org, Issue 9, No. 7; WEF (2006), *Competitiveness Report,* pp. 10–31; Federal Reserve Bank of Boston (2001), *Living Standards and Economic Growth: A Primer,* Boston, Massachusetts, pp. 3–6.
8. Brynjolfsson, E. and Hitt, L. (2000), "Beyond Computation: Information Technology, Organizational Transformation and Business Performance," *Journal of Economic Perspectives,* Vol. 14, No. 4 (Fall).
9. Holdren, J. P. (2001), "Energy Efficiency and Renewable Energy in the US Energy Future," Testimony for the Committee on Science, U.S. House of Representatives, February 28 (Washington, DC).
10. See Reference 7. In addition: Moskowitz, S. L. (2002). *Critical Advanced Materials Report: Final Draft,* pp. 4–40 (Prepared for: Virginia Center for Innovative Technology, Herndon, VA). Table also incorporates interviews conducted from 2001–2005 with executives, financial officers, and technical personnel within the advanced-materials sectors.
11. Ibid., pp. 10–15.
12. Ibid., pp. 20–22.
13. Ibid., pp. 18–20.
14. World Bank (2005), *Growth of Output,* World Development Indicators Data Base. For additional discussion of GDP trends between the United States and European Union, see Salvatore, D. (1999), *Europe's Structural and Competitiveness Problems and the Euro,* Presentation, EU Symposium 1999, Oxford, UK, Blackwell Publishers Ltd., pp. 1–4. Refer also to relevant data from the U.S. Bureau of Labor Statistics (BLS) and U.S. Bureau of Economic Analysis.
15. Bergstrom, F. and Gidehag, R. (2004), *EU Versus USA,* Timbro, (Stockholm, Sweden, p. 2).
16. Ibid., p. 6.
17. EurActiv.com (2005). "MIT Professor Challenges Perceptions of US vs. EU Productivity," *EU News and Policy Positions: Innovation and Jobs,* November 22. Note that the author of this article goes on to answer this important question by emphasizing the difference in the number of hours worked between European Union and U.S. labor, and what this difference says about different concepts and importance of leisure in European versus U.S. society. Although this is no doubt a factor, the bulk of the evidence points to other factors (e.g., technology and innovation) as more fundamental to these observed differences.
18. Aiginger, K. and Wieser, R. (2002), *Factors Explaining Differences in Productivity and*

Competitiveness, Austrian Institute of Economic Research (WIFO). Prepared for presentation at the Competitiveness Conference, January 30, 2002, Brussels, Belgium, p. 6.

19. Ibid. See also p.10, which shows the different productivity trends by country within the European Union.
20. European Commission (EC) (2002), *European Competitiveness Report: Staff Working Document,* Luxembourg, Office for Official Publications of the European Communities, p. 7.
21. Table constructed from information provided by the World Economic Forum (2006), *Global Competitiveness Report,* pp. 1–44. The table reflects U.S. leadership in technology and innovation and in business practices that make optimal use of harnessing these for productivity and economic growth. The table does not incorporate macroeconomic indicators, such as budget and trade deficits and surpluses, war expense, and so on, which arise from government policy and macroeconomic trends. When these are taken into account, we find that, from the years 2005 and 2006, the United States actually went from first to sixth position competitively in the world, with Switzerland and the Scandinavian countries moving ahead of the United States. Nevertheless, when we just consider technological and innovative output and business practices, the United States remains the foremost competitive force in the world. For the analysis of the U.S. competitive position, see Almeida, P. (1996) "Knowledge Solving by Foreign Multinationals: Patent Citation Analysis in the U.S. Semiconductor Industry," *Strategic Management Journal,* Vol. 17, pp. 155–165.
22. Jorgenson, D., Ho, M., and Stiroh, K. (2002), "Projecting Productivity Growth: Lessons from the U.S. Growth Resurgence," Federal Bank of Atlanta (Atlanta, GA): 1–12.
23. David, P. and Spence, M. (2007), Designing Institutional Infrastructure for E-Science, SIEPR Discussion Paper No. 07-23 (Stanford Institute for Economic Policy Research, Stanford, CA): 59–62.
24. Kim, D. J. and Kogut, B. (1996), "Technological Platforms and Diversification," *Organization Science,* Vol. 7, No. 3, p. 289.

Chapter 4

Facing Reality: The Risk Factor in Advanced-Materials Technology

To a great extent, the ability of a company, country, or region to excel economically and become and remain competitive depends on how they perceive and, from these perceptions, manage the set of risks that come into play in creating and adapting essential new technology. A number of studies show that different economies perceive and manage risk in different ways. The more successful either take routes to new technology that pose few risks or finds ways to manage risks so as to minimize or sidestep them, thus paving the way for companies to justify continued funding of cutting-edge projects. In contrast, less successful economies subjectively view risks as greater than they necessarily are, or have to be, and cannot find ways to effectively reduce the actual (or even perceived) level of risks involved.

A host of difficulties, some unexpected, often emerge when such new products begin to demonstrate commercial feasibility. Assuming that markets expand as predicted, possibly the most significant unknown is technical in nature. That is to say, will innovating firms be able to break through the various technical bottlenecks that must be faced with the scale-up and commercialization of new processes? Can the material be made with the requisite quality, in sufficient quantities, and at low per-unit costs so that it is competitive enough to topple the already existing, and known, technology from its market position?

And then there is the applications process. Will advanced-materials companies be able to successfully apply their technologies to the market industries, such as biomedical, automotive, electronics, and so forth? This depends in part on technical acumen. Success here also turns on the understanding of the immediate market and its requirements.

Even if the newcomer successfully handles the technical and applications issues, there is the question of the competition itself and what effect it might have. Typically, the existing players, sensing a competitive threat, move quickly to shore up their position. They adjust rapidly, and often successfully, to improve their products, reduce costs (and then prices), and in other ways search for all ways to strengthen their customer base and block, or at least weaken, the new competitive threat.

Pricing also does not simply fall into place according to some predetermined economic script. The pricing structure of a new material depends on a number of factors. This hinges on such things as the nature of the production process, the markets involved, and the competitive situation. The imposition of productivity-enhancing technology in the manufacturing process impinges directly on costs of production and, thus, on price structure. Beyond this consideration is how well the company determines what the market can (and will) bear. This last point, in particular, means that management's poor reading of the market and the firm's own limited capabilities can result in a misguided pricing scheme.

Weaknesses may also come to the surface in the management structure itself as a product moves into the commercial arena. A management team that functioned well during the research and development process may, for various reasons, falter in the market entrance and diffusion phases. A superb engineering team, for example, may not do so well when it comes to the practicalities of finding a particular market niche. The breadth and balance of a team's skill set becomes critical in new product development.

Finally, there are potential problems coming from the public policy and regulatory sector. Prior to commercialization, new products are under the radar screen, as it were, and little noticed by the public sector. But at the point when the product is on the verge of becoming a market reality, public policy issues begin to rise to the surface and may put barriers to commercialization and diffusion. We shall now consider these risks in more detail.

THE RISKS OF INNOVATION

Clearly, the United States has been able to take maximum practical advantage of the scientific and technical revolution that has led to new materials technology. Just as certainly, this has not been the case within Europe, with regard to results that penetrate deep within the European Union's economy. This study is not meant to be an exercise in technological determinism; quite the opposite. Even if technology is a central component in the lifeline of a nation' productivity—and we know from econometric studies hat this is indeed the case, certainly within recent decades—the success in developing a new product, process, or service hinges on just how well any particular innovation is managed, and this certainly turns on a nation's unique sociocultural roots and trajectories. We may consider such success in terms of risk. The process of innovation involves a sequence of risks along the route from creation to production to commercialization. How well (or poorly) an organization or even a nation manages those risks determines the degree of success (or failure) that can be claimed by that organization or nation. The focus here is on the most important sector of modern technology: advanced-materials creation, diffusion, and application.

Technical Risk: Can This Be Done?

Since the 1970s, countries with competitive advantages in new materials technology have been able to progressively develop and harness these advances for commercial ends. The particular relevance of these materials to semiconductor, electronic, and information-based systems, as well as energy and biotechnology, meant that these competitive nations were able to progress in their productive capabilities across a wide swath of the industrial landscape. Some material groups, such as engineering and electronic polymers, played a more central role in those decades, whereas others, such as nanotubes, have only in the more recent years been making their way onto society's technology map. But the ability of these materials to continue to evolve and confer economic blessings to those who attempt to develop, improve, and work them is less than clear. A major concern for advanced materials is accurate control of the structure and properties of these materials over time. In most of the important cases, the pertinent problems that need solving reside within the areas of engineering and practical process design, ranging from the need for rarified (but practical) tools such as scanning tunneling microscopes, capable of viewing and moving individual atoms, to large-scale mass-production systems.

Risks do not just involve the making of the materials themselves. In order for these materials to take on a commercial dimension, they must be applied to the various technologies and systems within the different industrial sectors. The degree to which advanced-materials producers work closely with these receptor industries, such as through technical services and joint research and development activities, determines the likely success of these application efforts. For example, severe bottlenecks plague the development of advanced (nano-) fibers. Nanofibers continue to resist attempts at easy solutions. This is due to portions of the billion nanotubes bundled together having hidden breaks, resulting in some fibers being perfectly straight while others are frayed and curling, thus weakening the final textile structure. In the case of nanocrystals, imperfections in the crystal structure reduce the efficiencies with which such materials convert solar energy into electrical power (only 3% compared with a 9% figure for today's solar technology). In a similar vein, the future prospect of bioengineered materials hinges on a number of factors. Most importantly, biorefineries need to achieve continuous, full-scale production to effectively compete in price against existing synthetic materials.

The following examples examine more closely the bottlenecks or technical risks with respect to both product synthesis and application faced by different advanced-material groups. These brief case studies discuss some of the past and ongoing problems faced by certain advanced materials. While not a comprehensive survey of such activity in the advanced-materials industry, these cases illustrate certain risks that are common to a wide range of new technological developments of the late twentieth and twenty-first centuries.

Thin Films [1]

By the beginning of the twenty-first century, thin-film technology proved itself a viable commercial entity, and it continues to expand into new markets. The United States leads other countries in the commercial realization of these processes and in the output of thin-film products and devices. Yet the range of application of thin-film technology still faces a number of technical and economic uncertainties when attempting to broaden the market base. One issue is capturing a continuous processing capability that allows the regular low-cost production of uniform and standardized thin-film products. A growing number of thin-film processes have emerged in laboratories throughout the United States and internationally. These include, for example, pulsed laser deposition, chemical vapor deposition, and electrostatic self-assembly. Improvements in these processes come from empirical work at universities, government laboratories, and industry. Increasingly, they surface from practical experimentation within small-firm facilities working to supply thin-film products to customers.

Difficulties abound also in attempting to apply thin-film technology to such complex industries as electronics. These include uncertain interface control; physical degradation of the polymer material in the presence of high temperatures, high electric fields, and exposure to solvents used in the circuit-printing process (which limits the types of circuits that can be designed); uncontrolled charge leakage between thin-film-based devices and circuit elements, resulting in lower operating life and increased signal interference; and reduced electrical performance and mechanical degradation due to impurities in the polymer or metal. Solutions in applications engineering of this type demands testing of different thin-film materials, components, and devices under different operating conditions within semiconductor elements and electronic systems. These "real-world" experiments require close, regular contact, and fluid lines of communication, between the thin-film manufacturer and the OEM within the electronic (and other) industries.

Advanced Fullerenes [2]

The original discovery of advanced fullerenes involved an understanding of bonding theory and structural analysis, as well as the use of complex and precise scientific instruments, such as the electron microscope. But once the initial materials had been captured, the emphasis in the field shifted to the engineering problems associated with practical production. Initially, chemical vapor deposition and catalytic processes, though potentially viable, produced significant amounts of undesired byproducts, for example, carbon black and amorphous carbon. The removal of these impurities from yields is expensive and limits the economic feasibility of these processes. Recent improved designs in such processes have come about through trial-and-error methods, such as altering operating conditions and assessing yields and economies for each set of physical and chemical variables.

Improvements in advanced-fullerene production center on two thermal processes. One technology applies an electric arc, using graphite to furnish the all-important carbon atoms. The major problem with the process is a practical, rather than purely "scientific," consideration: it is highly energy-intensive and is, therefore, expensive if used in mass production. An alternative approach, adapted from existing process technology used in metallurgical and power production, is the so-called "soot-flame" process. This technology creates commercial amounts of the more advanced (i.e., triatomic) fullerene materials by burning a mixture of acetylene (or related hydrocarbon) and the required metals to be "encaged." The advantage of the soot-flame process is that it is relatively cost efficient and production can be carefully controlled. The ultimate success of advanced fullerenes in the marketplace demands close coordination of their production to strict OEM specifications. This is so because the nature of the production process defines the types of fullerenes that emerge for consumer use. Increasingly, advanced-fullerene manufacturers experiment with altering process materials and conditions to supply the OEM market with custom-designed fullerene materials.

Nanotubes [3]

Not surprisingly, nanotube technology has much in common with advanced-fullerene development. In both cases scientific knowledge played a limited role in their discovery. Following the initial birth of the material, engineering skills and know-how took over as the prime movers of development and commercialization.

As with other advanced material categories, nanotube engineering offers an embarrassment of riches when it comes to the number of product and process possibilities. However, a persistent problem with nanotubes is that, at present, the existing processes cannot accurately control the structure and distribution of the product from one batch to the next. This limitation results in the output possessing a high degree of variability in the material's physical and electronic properties. As a result, older technology cannot satisfactorily custom design nanotubes for particular OEM applications.

There are two general methods of approach in the manufacture of nanotubes and nanoparticles. In the so-called "top-down approach," a bulk material gets, for lack of a better word, "chopped" up into nanosize bits; in the "bottom-up" method, molecules grow under controlled conditions (as in crystals) and then snap together into particular configurations, based on their charge and molecular chemistry. Though considered the most promising route, this latter process is far more complicated, and subject to all the laws of bonding that limit the ways atoms and molecules can be arranged. For example, it is a very difficult matter indeed to coax carbons to curl into a perfectly aligned tube rather than, as occurs too often, a thick, twisted scroll.

Most of all, the production of nanotubes is an engineering problem. It takes place using one of a number of processes including gas-phase catalysis, chemical vapor deposition, and laser-based technology. Each process brings benefits as well as dilemmas. Chemical vapor deposition involves heating a selected gas in a furnace and flowing the hot gas over a reactive metal surface. This route produces excellent yields with a low concentration of contaminants, but the resulting nanotubes contain a large number of defects, thus seriously limiting overall efficiency.

The catalytic process dictates that acetylene gas move over a catalyst located within a high-temperature furnace (approximately 700°C). In this approach, the acetylene molecules decompose and rearrange themselves into nanotubes. This method, which operates on a semicontinuous basis, can generate a significant amount of nanotubes. However, these are typically of the less pure variety and, therefore, are of limited use commercially. Another problem with this method is that metal particles from the catalyst tend to attach themselves to newly formed nanotubes. These particles magnetize the nanotubes thus limiting their use for applications in critical electronic components, such as transistors.

The laser approach for making purer nanotubes offers an alternative possibility, albeit with its own set of issues to be solved. The process involves the use of free-electron lasers (FELs). FELs operate at high energy levels and with very short pulses. They produce pure nanotubes by vaporizing graphite–catalyst mixtures. Removal of the impurities (such as spent catalyst and graphite materials), critical in the making of high-grade catalysts, for example, involves an inefficient, solution-based purification technology based on dissolution and precipitation of unwanted contaminants.

It is true that research is underway to develop other, radically new processes, such as improved electric arc technology. In 2002, IBM unveiled its new process for making single-walled nanotubes. The process involves a nanofabrication method centered on the silicon crystal technique. Although it promises minimum creation of byproducts and contaminants and little damage to the nanotube structures themselves, it remains under development with a successful outcome uncertain.

Beyond the difficulties that attend nanotube and nanoparticle creation is the equally thorny issue of application. Nanotubes may play an important role in the development of semiconductors, such as in the making of advanced sensor systems. The application of nanotubes here, however, has faced a unique set of problems. In addition to the issue of scaling up the processes to make the nanotubes and nanowires cheaply, much work still needs to be done on actually assembling them onto a microchip in a commercially feasible way. Once formed, the nanomaterials are harvested by being placed in a liquid solvent, such as ethanol, and blasted with ultrasonic waves to loosen them from the wafer surface. Researchers must then sort through the billions of nanowires or nanotubes to find the few that meet specifications they need for their semiconductor applications. Then too, orienting a nanowire onto a small 5-square-millimeter microchip remains at present very difficult, time-consuming, and costly process.

More recent research focuses on finding a way to produce nanomaterials separately, and then connecting them to form larger scale systems. In particular, University of California Berkeley researchers are attempting to grow the silicon wires and carbon nanotubes directly onto circuit boards by passing electric current through wires to specific locations on the microstructure where the nanotubes are to grow. In part, the economies from this process arise from the elimination of the cumbersome middle steps—the postassembly operations—in the manufacturing process, thereby opening doors to cheaper and faster commercialization of nanotechnology-based devices.

Despite these advances, the integration of nanotubes into such products as televisions and displays still face formidable difficulties. Indeed, products such as these are complex systems, so that altering one aspect of a technology will more than likely lead to other, interrelated problems that will need to be addressed. To take one example, bottlenecks continue to shadow attempts to integrate nanotubes applications in televisions, including controlling orientation and length and connections to silicon substrates and contacts, as well as high driver and peripheral equipment costs.

Advanced Ceramics [4]

Through the 1980s and 1990s, a key concern in the advanced ceramics sector has been to improve on the structural integrity of ceramic composite materials. An area of continued interest is to increase the strength and fracture toughness of the material, since sudden structural failure hinders wider applications of the material. The trick is to locate and select those few that have production and market viability. A promising category of structural composites still to be exploited is nanosilicon carbide–alumina and silicon nitride matrices. These composites possess a superior set of characteristics, including wear resistance, chemical inertness, anticorrosion properties, and excellent thermal insulation.

Also of vital interest in this area is development of new processes to make such composites as well as to manufacture existing nanoceramic materials more efficiently and with higher quality (e.g., improved compacting and pore size distribution). Different types of powder synthesis technologies, especially those based on thermal plasma and laser-based methods, promise much in this area, although complexities abound, particularly as relate to cost efficiencies.

Organic Polymer Electronics (OPEs) [5]

Electronic polymers have expanded their range of applicability over the last three decades. In particular, they extend the power and efficiencies of semiconductor systems. The next generation of such materials are the so-called organic polymer electronic materials (OPEs). The eventual widespread use of the more advanced forms of electronic polymers, including the OPE group of materials, depends primarily on the ability of the materials to overcome or sidestep a number of inherent technical (and economic) limitations. This is a critical point, as the current silicon-based systems may be approaching a technological barrier to Moore's Law. If such is the case, there will then be a heavy burden on electronic polymers to replace silicon.

The limitations that appear to be looming in the use of these potential substitutes for silicon are of more or less importance depending on the type of application. High on the list of known problems is the issue of uncertain interface control and the impact this lack of control has on electrical performance, such as capacities and efficiencies. Then too, physical degradation of the polymers in the presence of high temperatures, high electric fields, and exposure to solvents used in the circuit-printing process severely restricts the types of circuits that can be designed and reduces the overall life of such systems. Another difficulty that arises in the use of electronic polymers is uncontrolled charge leakage between devices and circuit elements, which leads to reduced noise immunity. A number of complications may result, but most important for many applications is the "bete noir" of wireless devices—signal interference over a range of applications. These internal issues also restrict information processing capabilities, a most serious difficulty. For instance, the speeds with which OPEs can process data are on the rise but still a long way from what crystalline silicon can provide. Crystalline silicon currently can achieve conduction speeds about 500 times faster than OPEs. Ultimately, the issue is one of quality or purity. OPEs tend to be less pure than silicon, resulting in reduced electrical performance and mechanical degradation. Purification by some chemical or physical technique can be a difficult and expensive option.*

Attempts to apply OPEs to electronics pose their own problems. A barrier to applying OPEs to technology, such as in radio frequency identification (RFID) systems, centers on the printing process. As the tag must be smaller and have greater ability to process information, the printing of the circuit onto it has had to be very fine. The most effective printing technique is "screen printing," which has not been able to print precisely enough to make OPEs effective for electronic applications. The development of new and improved screen-printing processes may offer a solution. In addition to improvements in screen printing, advanced versions of alternate printing processes are being developed, such as "stamp" printing. Similar to OPEs in displays, there are issues of variable consistency and uniformity as well as degradation problems under high temperatures, in strong electric fields, in high humidity conditions, and in the presence of solvents used in the circuit-printing process.

These case examples beg the question as to what can reduce the technical risks involved in developing and applying new productivity-enhancing innovations within industrialized countries. Is scientific advance, as is commonly believed, the key? All too often, mention is made of the so-called "science-based" industries, such as electronics, biotechnology, and advanced materials. But to what extent is science, and especially cutting-edge science, the gold standard requirement of modern technology?

There is, in fact, a rich (and growing) literature coming out of such disciplines as economics and the history of technology that claims that theoretical science on the one hand and engineering practice on the other are two distinct fields—two separate cultures, really—with their own goals, values, language, and problem-solving methods. Whereas the pure scientist strives to increase knowledge of the universe for its own sake, the practical engineer looks to solve particular real-world problems in order to "make things work." And whereas the scientist employs abstract mathematical reasoning and gains professional status by publishing theoretical articles in peer-reviewed journals, the working engineer works with "practical" mathematical constructs (e.g.,

*Experts in the field do not all agree on this point. Some maintain that OPE production would not necessarily involve expensive purification technology because OPE materials work quite well without needing to be very pure at all. They claim that this is, in fact, one of the big advantages of OPEs over silicon. Also, energy utilization may favor OPEs since they are synthesized at room temperature, in contrast to the high-energy requirements associated with silicon-based production.

approximation methods, dimensional analysis) and empirical experimental techniques, and secures professional stature by patenting inventions, saving employers time and money, and successfully guiding new products into the marketplace.

Certainly, the examples above do not seem to point to advanced science as a major driver of new materials innovation. The major problems encountered in our case studies have less to do with an understanding of atomic and molecular structures and interactions, although there is some of that during the phase of initial research and discovery, and far more to do with what we might consider practical engineering concerns, such as identifying all possible advanced-material technologies worthy of further development, hitting upon the appropriate set of operating conditions, using empirical testing, designing continuous operating processes, properly incorporating new materials into OEM and consumer products, and so forth.

Interviews conducted with technical personnel and engineers working in these various advanced-material fields, along with data and information from case study sources, confirm this assessment. Table 4.1 assigns a number value from one (least important) to ten (most important) for various categories of problems encountered in the development, commercialization, and application of each of the major advanced-material groups. It is understood that each one of these product life-cycle

Table 4.1. Technical bottlenecks to the production, diffusion, and application of advanced-material technology

Technology	Scientific theory	Scientific instruments	Scientific data (empirical)	Operating conditions	Physical capital design and availability	Continuous design processing	Interindustry linkages and communication	Selection of viable candidates from large pool of possible technologies
Nanotubes	5	6	7	9	9	9	9	9
Advanced fullerenes	4	6	8	9	9	9	9	9
OPEs	4	5	8	6	9	9	9	8
Engineering polymers	2	3	7	6	8	9	7	7
Electronic polymers	3	4	8	7	8	9	9	9
Thin films	5	7	9	8	8	9	9	9
Biorefinery products	3	5	9	9	9	9	8	6
Nanospheres	5	7	8	9	8	8	9	9
Nanocrystals	5	8	7	9	8	8	9	9
"Super alloys"	2	3	6	9	9	9	9	9
Advanced ceramics	2	3	6	8	9	9	9	8
Average	3.6	5.2	7.5	8.1	8.5	8.8	8.7	8.4

Scale: 1 = least important, 10 = most important.
Source: [6].

stages would necessarily register different "relevance" profiles. For example, the development stage might have more use for fundamental science, whereas the applications phase might require practical engineering. The table shows the average importance of different "relevance" categories over the life cycle of a product group and so answers the question, "what is the importance of science or engineering in creating a product and getting it into the marketplace so that it has an economic impact in society?" The first two columns in the table refer most directly to scientific issues. As one proceeds from right to left, columns relate more to empirical (non-theoretical) and practical engineering concerns. The last row of the table provides an average score for each of the "relevance" categories. As is clear, average values tend to be much higher once we get past the first two or three categories. When science does come into play, theory plays a decidedly secondary role compared to the more "practical" use of scientific instruments and empirical data. This result makes sense because the latter two provide knowledge inputs potentially useful in engineering design. The remaining categories relate to candidate selection, process design, and market application of the advanced material products.

The Financial Capital Dilemma: Can This Be Paid For?

The technical hurdles described above offer significant challenges to the new-materials industry. These barriers cannot be breached without what have always been two important components of technological innovation: financial and human capital. Any new industry that evolves around new technology faces a particularly difficult situation in both regards. This is especially true of the new materials industry.

The signal importance of process design, equipment, and instrumentation underscores the risks resulting from shortages in financial capital with which to conduct R&D, scale up production, and plan market-entrance strategies. Particularly onerous, especially for smaller firms, are the incessant demands of fixed costs that keep the business running. These include not only infrastructural costs but labor as well. Generally considered to be a variable cost, labor operates as a constant in the case of high-technology start-ups, at least up to the point when production begins to take over, since the same core of scientists, engineers, and technicians work to keep the company afloat as they develop new products and processes for market. Purchase of equipment and machinery also consumes vast amounts of financial capital. As firms work to scale up, they must invest in more and larger equipment and mass production systems, thus ratcheting up capital consumption many fold.

During America's first materials revolution from the 1930s to 1960s, internally generated and institutional capital supplied the greater part of the money requirements of corporations for R&D and postdevelopment placement of innovations in society. Companies such as DuPont and Jersey Standard (Exxon) were quite capable of funding themselves out of retained earnings. As established, diversified (or at least multiproduct) companies with proven track records, they also accessed capital in the form of long-term loans from institutional banks. These companies tapped into these latter pools of capital only in the face of the larger and more expensive projects. The government also served as a source of funding, mainly in the years leading up to and during World War II. Following the War, materials innovation once again turned to retained earnings of the large and increasingly multinational corporations.

By the 1980s, the innovative start-up firms, often licensing and then working with promising patents from universities and government laboratories for commercial development, became the sources and stewards of society's newest and most advanced technology. The rise of these firms as innovators meant a shift in those sources of capital for R&D. These firms could not depend on institutional capital, which shied away from these more risky ventures. Nor could they turn to retained earnings to fund commercialization, since their profit margins tended to be small and, as often as not, they operated in the red for the first few years of their existence while they waited for the placement of their first products into the marketplace. Two major sources of funding kept these high-risk companies solvent as they developed their first new products. Government monies in the form of grants from such agencies as the Departments of Defense, Commerce, and Energy, and the Environmental Protection Agency (EPA) proved to be (and continue to be) especially important sources of development capital. The government's interest in commercializing federally funded products and processes for the general good of society motivates many of these grant opportunities. The second and most important form of capital has been venture capital. Especially in the United States, the flourishing of the venture capital industry played a critically important role in supporting the risk-driven new-technology start-ups (see Table 4.2).

One of the major risks for would-be innovators in new materials is locating and attracting appropriate, sympathetic, and patient venture capital groups. During periods of economic retrenchment, such as in the years of recession from 2000–2003, the willingness of the venture capital community to risk its capital in funding new technology was sorely tested. The dot-com collapse during these years brought home to the venture capitalists in a very palpable way that even those "successful" technologies once cherished by the venture capitalists had their Achilles heels, and had to be closely watched in the future. Such failures quite reasonably made the more conservative portion of that community that much more reluctant to invest in the new wave of technology.

Table 4.2. Distribution of funding sources for advanced-material R&D, United States

Source	1970	1980	1990	2000	2010	2020
Government	18%	17%	15%	12%	10%	10%
Institutional	2	3	3	4	5	5
Retained earnings	3	3	2	2	2	2
Venture capital	35	38	55	65	70	75
Other	42	39	25	17	13	8

Source: [7].

The Human Capital Dilemma: Can This Be Staffed?

Equally problematic is question of the availability of appropriate human capital. This issue depends not only on the aggregate numbers of personnel within a country, but also the elasticity or mobility of labor.

As previously noted, it is questionable whether an abundance of first-rate theoretical scientists working in the advanced materials field can propel the technology forward. Closer to the mark are industrial scientists or, even more relevant, those with engineering knowledge and firsthand experience with the technology. In either case, it appears that a shortfall exists between the continual demand for advanced-materials specialists and the existing pool of relevant talent. In the United States and Europe, there has been a significant fall in the number of engineering graduates, at least since the 1980s. The economy, within the United States and globally, simply did not cry out for such talent during these decades, as it had, say, in the two decades following the Second World War. Within the United States, the collapse of the space program in the 1970s and the fall of the Berlin Wall in the late 1980s put a severe cap on the number of scientists and engineers in demand by the government. The Vietnam War and the cultural changes in society attached to that trauma that have extended into the twenty-first century further eroded the former luster of such careers. As a result, over the last thirty years, not only has the number of such professionals declined, but the mathematical and scientific ability of Western youth, especially in the United States, has seriously decayed.

But it is not enough to say that a shortage of practicing engineers poses the most serious human capital risk for a nation's advanced-technology potential, for even as the United States experienced its engineer shortage, it has excelled in advanced-materials development. Over the past two decades, economists and historians have examined the role of the so-called "gatekeepers" in creating new technology and harnessing it to the engine of economic growth. These vital agents of innovation, because of their interests, skills, and training, work easily within and between the technical, financial, market, and even political arenas. This ability is important because modern technological growth is multidimensional in nature, in that it incorporates all these components simultaneously and within a closely interactive network context. Gatekeepers bring greater competitiveness to the organizations within which they work because they excel at rapidly bringing together and tying these various disciplinary strands together into a coherent, integrated, and directed whole. These individuals play such a key role today because larger firms that traditionally led in new technology creation tend to become rigidified and less responsive to innovative ideas. Their multilayered departmental and divisional organizational structures create informational barriers between disciplines, thus slowing down and even stopping cold the innovative impulse.

Table 4.3 shows the results of a survey of 50 successful high-technology firms, most of them involved in one way or another in the development of advanced-material products. The table tells us what percentage of these firms succeeded primarily

Table 4.3. Human capital requirements for successful advanced-materials firms: 1990–2004*

Profession	%
Theoretical scientist	0
Industrial scientist	1
Engineer	1
Venture capitalist	2
Business executive	3
Entrepreneur	3
Government official	0
"Gatekeeper": team	30
"Gatekeeper": individual	60

*Percentage of firms within sample of 50 firms that succeeded primarily because of profession indicated.
Source: [8].

because of the actions of each of the professional groups listed: theoretical scientists, engineers, and so forth. The last two categories distinguish between multidisciplinary individuals and the effective team composed of individuals who specialize in different areas but who effectively coordinate these disciplines into a closely integrated unit. The latter "coherent team" acts as a de facto gatekeeper within the organization. The striking conclusion from the table is that the vast majority (90%) of successful firms must operate in an integrated multidisciplinary environment. No one discipline, not even engineering or venture capital, despite their importance, can "go it" alone. They must be linked within a network of interconnected and mutually reinforcing disciplines.

In this light, the issue of human capital supply takes on a different meaning than is traditionally understood. An adequate supply of practicing engineers or venture capitalists, while certainly important, will not by themselves lead to competitive advantage in new technology. What is essential is a reasonable number of strategically placed gatekeepers (or gatekeeper teams). A successful gatekeeper, for example, may have started out his or her undergraduate schooling in engineering but decided to finish up as a business or finance major and then went on to pursue an MBA. Or he or she may be majoring in management but minoring in engineering. Although we would not count this person as a graduate engineer, he or she would have certainly obtained a technical training that could be combined with finance, marketing, and so forth. This multifaceted individual would then play a major part in pushing a firm in new and profitable directions. The risk factor here is that a country or region might not be able to produce individuals at the right time and place who are capable of operating in a multienvironment context. Historical and cultural factors play critical roles in whether a nation can create the requisite conditions for the creation of a gatekeeper mentality.

The "Selection" Dilemma: Can We Pick the Winner?

The technical accomplishment of a company or nation is often linked in discussions to patent output, and rightly so. The quantity of patents generated does indicate the level of activity geared toward the practical pursuit of commercial products and processes, as opposed to purely scientific pursuits and the focus on knowledge expansion for its own sake. Nevertheless, most patents never enter the market system. This is so for a variety of reasons. Most important is the fact that, although the nature of the innovation in question may satisfy the criteria for the granting of the patent—criteria that can differ significantly from one country to the next—it may not be such that it can be commercially realized, either because of product quality or process inefficiencies. Indeed, the requirements of the latter are generally much stiffer than for the former, as successful entrance into the market demands understanding and satisfying a great number of real-world variables that may not be thought through completely in a patent application. One example of this might include reduction of price to competitive levels through a myriad of minor, non-patentable improvements in the manufacturing process. As Table 4.4 shows, the number of patents in the advanced materials field has been growing at a rapid pace, and this trend is expected to pick up after 2010. Although the percentage of these patents that have a commercial significance is on the rise, it is clear that the far greater proportion of these patents will never become powerful, or even active, players in the economy.

This gap between a patent on an innovation and the actual ability of that technology to enter and impact the market creates a serious risk for both companies and their home countries. In this riot of possibility, to what extent a company and a nation can pinpoint just those ideas and concepts that are the most likely to be commercial successes and push the limits of economic growth is a critical question. National and regional competitive advantage can very well rely on this insight to sort the wheat from the chaff. Not being able to do so depletes valuable resources and wastes precious time on dead ends. In the meantime, competitors savvier and, therefore, more efficient in selecting the commercially robust patents to license and develop can focus a greater concentration of resources on the most viable candidates. Thus, they conserve personnel, money, and materials on those routes most likely to meet economic and societal goals and so achieve economies of specialization. Fur-

Table 4.4. Patent trends and the percentage of patents that enter the market—advanced materials: 1970–2020

	1970	1980	1990	2000	2010	2020
Number of patents	500	2,000	3,500	5,750	8,500	12,500
Percentage of patents that enter the market	2%	4%	6%	8%	10%	12%

Source: [9].

ther, they can get these technologies to the market faster and so realize all the benefits that accrue from a first-mover advantage.

The Market Dilemma: Can We Sell This?

Assuming that those industries that consume advanced materials expand as expected—an assumption that has its own risks—the market has posed, and continues to pose, three major risks for advanced technology: (1) competition from other materials (both existing and new-material technologies); (2) crowding of a particular material field, leading to dangerous overproduction; and (3) public perception risks. Those companies and nations that successfully bring new materials to market effectively negotiate these risks. They, or others, must handle these same risks over the next few years if they hope to shepherd ever newer technologies into the economy.

The Substitution Factor

There is, of course, competition between the various new materials. But possibly more critical is the hard reality that new materials must first unseat existing, tested, and well-known older material technologies. The very fact that consumers already know a material creates a difficult hurdle for the newcomers to overcome. Then too, the extant and rapidly maturing industry, feeling the hot breath of competition on its neck, strives, often effectively, to improve its product in order to retain its markets.

Advanced materials face similar and quite formidable resistance in their bid to push existing technology to the side, more so now than in the past. From the 1930s through the 1950s, new materials such as nylon, polyethylene, and synthetic rubber quickly found their markets, in part because they were introduced and used during wartime, and also because these new technologies were clearly superior (technically and economically) to what was available at the time. The older materials, such as rayon, simply could not even begin to compete with these new synthetics in the mass market, no matter what strategies they might grasp onto in order to retain their customer base. At best, they remained a familiar presence in certain niche markets or with those consumers who were skittish in the face of the new. But soon, with lower prices and superior properties enticing even the most wary, the new materials won over even the most reluctant.

But can we not claim a similar inevitable dynamic for today's materials technology? The answer is not quite. Today, competition is more severe than in the past. In fact, those very same advantages claimed by the fibers, plastics, and synthetic resins that first emerged in the World War II period and that so easily whisked away the existing competition at that time, appear to have a durableness to them that protects their industries from today's upstarts. Of course, no global conflict on the order of the Second World War has come into play to accelerate the use of the newest materials. More fundamentally, an existing and familiar materials technology is a formidable force. It is also a highly malleable one, allowing of multiple manipulations of the macromolecular structures to improve upon as well as extend possible properties and manufacturing processes. In the face of the powerful presence and ever changing, and ever improving, face of the older materials, the newcomers confront greater barriers than did their counterparts when first introduced decades ago.

Market power does not only rest with the organic macromolecules. Existing inorganic materials also claim a tight grip over their markets and the new generation of materials face a difficult time in ousting them from their exalted perches. As previously noted, one of the most important technological battles that is close at hand pitches the ever-present silicon against the rising tide of the organic polymer electronic (OPE) materials [10]. There are a number of reasons why silicon can prevail, at least in the near future. Crystalline silicon is a superb material from a technical point of view to use in current applications. In particular, the speeds with which OPEs can process data are on the rise but still a long way from those of crystalline silicon. Currently, advanced materials, such as OPEs, cost more than silicon to produce. This may change as new materials are manufactured in greater quantities using mass production designs, but it is not clear that the price differential will change much before 2015 to alter the equation significantly. And then there is the entrenchment of the silicon suppliers. This will be difficult to break since these suppliers currently enjoy close and long-term relationships with customers. Moreover, the established silicon-based chip technologies continue to improve performance and design. A good example of this is the current developments in the area of antenna technology. These new and improved systems, such as the printed antenna technique, employ passive organics as substrates but use metal (e.g., silver) as the conductive element. This technique may be able to reduce the cost of the critical antenna-die systems without the need for active OPEs. The silicon chip industry is also striving to remain viable by improving one of its most critical steps with new and advanced forms of lithography, such as ultraviolet, optical, and electron-beam lithographic systems.

Similarly, in displays, silicon is the material of choice in thin-film transistors (TFT) and used in flat panel displays incorporated into laptops. In this case as well, OPEs as a material for displays would have difficulty competing with silicon and other display materials as these are so solidly entrenched. In addition to silicon and more traditional (nonconducting) organics, a display depends on its major component, the liquid crystal substance. There is no universal agreement as to the degree to which OPEs can practically substitute for this material in LCDs. There appear to be limitations in the potential use of OPE materials in general. The liquid crystal, as used today, is composed of highly polar molecules that are vital to the working of the display, a property not shared by OPEs.

Other advanced materials face similar resistance in their bid to push existing technology to the side, more so now than in the past. The future of bioengineered materials hinges on a number of factors. Most importantly, biorefineries need to achieve continuous, full-scale production to effectively compete in price against existing synthetic materials. The companies involved in the technology must accelerate their technical services programs in order to locate and capture increased market share. This depends as well on finding new applications for the materials. Market risks associated with advanced materials may be reduced by combining with existing devices and systems.

A promising route for the newest materials is as supplements to older technology. In this way, the OEM market for established petrochemicals can serve as an important source of demand for the new generation of advanced materials. For example, nanomaterials themselves may extend the life of silicon chips. Harvard University, for example, is examining the use of phosphide nanowires, and the United States and other countries (including the Netherlands) are exploring the use of carbon nanotubes to build basic logical circuits. In this way, the life of the silicon chip industry may last 10 years or longer due to new product designs and manufacturing processes. In particular, NASA's Ames Research Center has investigated nantubes to replace copper conductors to interconnect parts within integrated silicon-based circuits. These nanotube interconnects can conduct very high currents without any deterioration (currently a problem with the copper interconnects used today). The process involves "growing" microscopic, whisker-like carbon nanotubes on the surface of a silicon wafer through a chemical process. Researchers deposit a layer of silica over the nanotubes grown on the chip to fill the spaces between the tubes. Then the surface is polished flat. More multiple, cake-like layers with vertical carbon nanotube "wires" can be built to interconnect layers of electronics that make up the chip.

The Price Factor

Will the advanced-materials industry be able to meet the pricing requirements of an expanding market? Part of the issue is technical, part economic, and part related to management. Certainly, only when the production and application of advanced materials is on a mass production basis will prices come down. This is the classical model of price–demand relations, where lower production costs allow reduced unit prices which, in turn, increases the scope of demand, further inducing expansion in mass production improvements, and so forth. As the price continues to fall, additional markets come into view.

Pricing judgment is of critical concern for it determines rate of substitution. Prices must come down dramatically and in a relatively short time if the new material can hope to dislodge existing and known technology from its entrenched position. When prices are high and difficult to pressure downward, the chances of successful replacement of a new material for the existing technology sharply diminishes. Also, effective competition from impacted technologies against the new and upstart material seriously undermines the prospects of even the most promising innovation.

Table 4.5 demonstrates the close relationship that exists between the price of advanced-material technology (in this case, nanotubes) and the range of accessible markets. As the table suggests, setting the appropriate price at the time of first market entrance is critical, as it determines the diffusion rate of the technology. The ability of the management team to measure and interpret market demand informs their pricing decisions, given the state of their production technology.

The Competitive Factor: Rivalry, New Entrants, and Overcapacity Risk

New technology fields run the risk of overcapacity and reduced profits. The critical question, then, is whether overcrowding in the advanced-materials industry can erode future profit margins within companies, thus hindering their ability to support critical R&D? This overcapacity comes about through the licensing of too many firms and by other companies finding ways around patent infringement and developing their own processes. For instance, by the 1970s, only three major producers—GE, Bayer, and Dow—competed worldwide in the polycarbonate synthetics. But since then, aided by engineering firms designing turnkey operations and the force of globalization, many more producers have sprouted up, especially in Asia, resulting in overcapacity in the industry.

More intense rivalry and increased entrants in an industry are often associated with an increase in patents disputes, which represents yet another market-related risk, especially for the smaller, innovative start-up firm. The rate of patent filing within the United States and internationally, and the scientific complexity of many of the patents, tends to trigger a flurry of patent infringement cases that slows R&D and commercialization. In the economic boom of the 1990s, the filing of new patent lawsuits increased rapidly. The filing of intellectual property legal cases peaked nationally in 2000. Although this number has declined since then, both in terms of absolute numbers and as a percentage of total new cases filed, the level remains higher than in the mid 1990s. In 2000, there were nearly 9,000 new intellectual property cases filed nationwide, representing 3.4% of the total cases that year. By 2003, this figure, though declining somewhat to 8,254 (accounting for 3% of total lawsuits filed that year), still stood significantly above the levels of the early 1990s. This means that corporate resources that might otherwise be used for technological development must be diverted to defending or pursuing these cases. Indeed, with more money on the line riding on the outcome of patent lawsuits, companies plow more of their retained earning in these legal battles. The

Table 4.5. Price–market relationship for nanotubes

Price per pound	Market
>$15,000	Research material; limited commercial applications (e.g., microscope probe tips and specialized membranes)
$15,000	Flat panel displays for PCs; TV sets
$10,000	Microwave devices (e.g., antennas); radar-absorbing coatings for aircraft
$200	Fuel cells; batteries
$100	Drug delivery systems; commercial composites for fabrics, beams, structural members
<$100	Electronic devices; lightweight automotive and aerospace components

Source: [11].

median cost today of litigating a single patent case is between $2 million to $4 million [12].

In this intensely competitive (not to mention litigious) environment, the smaller firm, totally committed to developing and introducing the contended technology to market, lacks the financial and manpower resources to defend its patent claim in court. Further, it appears that, using the percentage of patents litigated to date, litigation rates in the United States are highest in those fields most critical to advanced materials, including electronics, biotechnology, and the mechanical and electrical industries. Together, these fields account for approximately 70% of all patents litigated in U.S. courts.

The Management Dilemma: Can Our Team be Effective?

The risks involving management are twofold. First, there is the risk that, as the new technology evolves over the course of its product life cycle, the management team will lose touch with new requirements and criteria for success. For instance, as the new technology begins its course through the market, the firm must be able to read signals from users as to what parts of the technology must be modified to break through market resistance. The second concern is that the organization and its management, for various reasons, may weaken in their resolve to commit the time, costs, and resources required to even begin exploring the newest and most promising technology. These two management-related risks combine in a potent manner to cause the growing firms to perceive a high level of risk in continuing to pursue innovation, thus reducing their calculated desire to do so.

The signal issue with respect to our first concern is whether the firm, which appeared so competent in developing a new product or process and bringing it to the market's edge, will falter in entering and diffusing into those markets. Certainly, significant differences exist in the capabilities needed in the earlier and later phases of a product's life cycle. Typically, as the life cycle proceeds, the talents needed shift from the purely scientific and technical to expertise in marketing, pricing, servicing, and so forth. The matrix in Table 4.6 shows how management requirements for pushing forward commercially a typical advanced-material technology shift over time as product development moves through its different phases, from original research (Phase 1) to market placement (Phase 5)

But it is far from clear that advanced-materials firms worldwide can adjust rapidly enough in their management structure to the requirements of the advanced-material life cycle. In the first place, as Table 4.7 below shows, although the costs and uncertainty of success for market placement of new technology has increased, the life cycle of those technologies that "make it," and advanced materials specifically, have become more compressed over the last 20 years, in large part because of the technically dynamic situation in such industries as electronics, communications, and biotechnology.

As the cycle has shortened, it has become ever more important for firms to be malleable in order to quickly modify their management capabilities from the initial

Table 4.6. Shifting management requirements of an evolving advanced-materials technology

	Product Life Cycle				
Capabilities	Phase 1	Phase 2	Phase 3	Phase 4	Phase 5
Scientific	4	2	1	1	1
Technical	6	8	9	8	7
Financial	5	5	8	9	10
Marketing	2	3	5	9	10

Scale: 1 = Not Important, 10 = Very Important.
Source: [13].

investment phase to the downstream commercialization stage. More so than in the past, market dynamics may make a new technology obsolete, or at the least less attractive than anticipated, unless management closely observes and responds to trends. But as a firm grows and becomes ever more hierarchical in nature, our second concern comes to the fore as well-known rigidifications in the organization can severely hinder intracorporate information flow, and in turn, organizational flexibility in addressing changing competitive conditions. The compressed life cycle induces improvements in older technology and also the entrance of newcomers into the field. Even if a firm succeeds in shepherding a new-material technology to market, the threats of patent disputes and overcapacity loom. Thus, innovation appears to be a far more risky endeavor than in the past, a perception that is heightened as the costs for bringing a new product to market have steeply grown, especially over the last half century.

As the costs and time factors increase the risks, managers face the decision of continuing with development and commercialization of new technology. Technology promises great rewards for a company if it proceeds with innovation, and potential disaster if it does not, in part because competitors in advanced technology are likely to offer it sooner rather than later. Moreover, the more time a firm spends on developing a new technology, the more likely it is to continue, since resources have already been sunk into the project. On the other hand, if management perceives

Table 4.7. Lifecycle trends in advanced technology

Year	Total life cycle (years)
1930	15
1950	12
1970	10
1990	8
2000	6
2010	5

Source: [14].

risks growing in terms of resources spent, and if the company believes it can only achieve a viable technology at tremendous cost, it may abandon the effort so as not to throw good money after bad, and fall back on simply improving existing products and processes as their competitive strategy.

The case of engineering polymers provides a sobering example of how once mighty R&D organizations may reach a point at which they abandon pursuing new technology. Perceived risks play the dominant role here. The larger companies in the field want to make the newly acquired materials by incurring only incremental costs. As one recent study on R&D in the chemical industry concludes:

> Evidence suggests the presence of certain complacency, and perhaps even disillusionment with investment in innovation in the chemical industry. . . . [Accordingly, the] industry [is] lagging behind such innovative sectors as electronics, pharmaceuticals, and even oil and gas. [15]

Since the late 1980s, the industry sees innovation and technology creation as too risky, especially in the commercialization of really new materials, such as advanced polymers (e.g., Shell abandoned development of the once-promising "Carilon" polymers).

These risks further support the important role played by the gatekeeper in advanced-materials progress. The gatekeeper's ability to see at all times beyond the narrow confines of a single field or specialty imparts an exceptional flexibility that allows him or her to rapidly and efficiently change hats, as it were, as the product rapidly evolves through its life cycle. We will explore in more detail the growing role of the gatekeeper in modern materials innovation within a cluster context in later chapters.

Public Policy and Perception Dilemmas: Will Government Action Derail Our Strategy?

Public policies and perceptions regarding technology also pose risks to countries and regions attempting to boost productivity and competitiveness through innovation. Social, cultural, and even political factors play a significant role in accounting for the disposition of the marketplace for technological development. In the late nineteenth century, for instance, the German government actively sought, through public policy, to create markets for its burgeoning coal-tar chemical industry. Similarly, at the end of World War I, having witnessed the strategic importance of chemical research for national defense and self-sufficiency, the U.S. government worked with chemical trade groups to publicize the good works and economic potential of the chemical industry. This campaign resulted in a more favorable public perception about the industry. This development, in turn, created and supported markets for chemicals and new materials within the United States and disposed potential investors more favorably toward this sector. The surge of American chemicals and the first advanced-materials revolution within the United States flowered within this favorable climate.

On the other hand, public policy may dampen technological progress and its benefits to society through onerous regulations and misdirection of resources. In this sense, government action, depending on the direction in which it is applied, may pose a formidable risk to a nation's continued progress and competitiveness. Trade regulations that impose barriers to new technology and limit healthy competition create a poor climate for innovation. Environmental regulations, if loosely drafted and too costly for industry, create a severe burden for small- and medium-sized firms (SMEs) which bear so much of the weight of technical innovation today. Government programs that direct most funding to "big science" projects miss the mark in developing practical technology that can create economic growth.

A key issue in considering the role of government in advanced-materials creation and diffusion is the type of government involved. World War II signaled the importance of central or federal governments in the funding and guiding of science and technology. As war approached, both the Allied and Axis powers, appreciating the strategic importance of technical innovation, centralized the control of the resources required for fashioning new materials and their associated technologies. Success rapidly followed as nylon, synthetic rubber, aviation fuel, explosives, and advanced alloys entered into war. The development of the atomic bomb represents a technological climacteric to this wave of creative destruction. Following the war, all these technologies entered into the mainstream economy, first in the United States and, by the early 1950s, Europe and Asia. Through the postwar period of the 1940s and 1950s, much of the new material-based technologies could be traced, either directly or indirectly to past and current funding and support from federal governments, both within the United States and Europe. This was, in fact, the model put forth after 1945 by Vannevar Bush—industry desired short-term results and so could not be relied on to conduct basic research. Thus, governments must fund universities, which support the free market in ideas, even without commercial application. After 1958, the year of passage of the National Defense Education Act, federal money earmarked for scientific research poured into U.S. colleges for graduate scholarships, laboratories, and equipment. By 1964, the U.S. federal government funded nearly two-thirds of total domestic R&D. By the late 1950s and through the 1960s, the hand of government continued to be felt in materials research, especially within the United States, with the launching of Sputnik by Russia (1957) and the subsequent ratcheting up of America's space program.

A dramatic shift in this pattern of federal support for advanced-materials technology occurred in the 1970s, particularly within the United States. The Vietnam War diverted funds from the development of peacetime technology. Although the war did spawn new-materials technology, such products as Agent Orange and Napalm, both products of the chemical industry, did not translate readily into commercial technology. From this point, new-material technology did not rely as heavily on federal support and influence beyond the sporadic infusion of monies into newer firms that won relatively small grants from programs sponsored by the Departments of Defense (DOD), Energy (DoE), and Commerce (DoC). Since the 1970s, local departments and agencies have taken over the federal government's role in advancing new-materials innovation (and technological development in general). The growing sense of competition for economic power, as reflected in employment and an expanding tax base, between states, counties, and cities and towns fueled a grassroots movement within localities to support new-technology creation. The rise of industrial and technology centers and clusters through local funding, zoning, taxation, and other incentives became an increasingly important source of new technology. This development has also meant a bigger role for state and local political activity in the innovation process. For its part, central governments within the United States and Europe remain wedded to supporting the "big science" concepts and the R&D programs of large corporations. But this stubborn adherence to past models misses the point of modern day innovation, which, more than any time in the past, must be guided by universities, incubators, and start-up firms that reside close to one another and nurture empirically created technology. They are linked to the immediate marketplace through a network of contacts, based on practical "down-to-earth" engineering, and are an integral part of a cluster of interrelated firms located close together within an area in order to capture critical economies in the flow and transfer of knowledge, skills, and technology.

The technical, resource, market, economic, and public policy risks that we have discussed in this chapter provide a framework for understanding the great divide that appears to separate the United States and Europe with respect to the creation, diffusion, and adoption of the new-materials technology and, in turn, economic progress and regional competitive advantage. Table 4.8 reviews the various risks that potential entrants into the advanced-materials field must encounter, address, and overcome. The table shows the general "risk category" on the left and, within each category, particular type of risk on the right. In each case, the first entry listed under each "risk type" would be the risk factor; the second entry (to the right of the "vs") indicates the optimal possibility. For example, an important technical risk is that a company focuses too heavily on science and theory when it should be gaining a practical engineering capability. Under the "financial capital" category, an economy that offers only an institutional investment route for business development at the expense of a robust venture capital community will likely suffer in the advanced-technology field since investment capital, which is by nature conservative in its ways, tends to flow to the less risky options. Note that under the category "government," the risk is that the federal agencies, rather than localities and start-ups, control the innovation process. The last risk category in the table, "public perception," refers to the degree of acceptance within a country or region of new technology. This is a function of a country's culture and history. The recent trade dispute between the United States and the European Union over genetically modified foods (GMFs) is a case in point. We will have more to say on this topic in a subsequent chapter.

These risk categories and types are independent and self-contained, but also closely interrelated and mutually reinforcing. The degree of substitutability of a new product for the old, for instance, may depend on the technical quality of the material as well as on the newcomer being able to enter the market at a lower price than its rival. This latter ability also may involve a technical component in the form of a mass production process. In another type of interconnection, the venture capital

Table 4.8. Risks in advanced-material technology

Risk category	Risk type
Technical	Theoretical science vs. Practical engineering
Financial capital	Investment vs. Venture
Human capital/management	Specialists vs. Gatekeepers
Market	Distance from market vs. Close links to market
Substitutability	Low degree vs. High degree of substitutability
Prices	High costs (prices) vs. Low costs (prices)
Government	Federal vs. Local
Government	Federal (+ large corporate subcontractors) vs. Industry (small start-ups)
Public	Negative vs. Positive perceptions

community ("financial capital"), in continual search for new technology that will find its demand, selects those projects to develop that they feel have close linkages to markets and, in fact, the venture capitalist often helps to establish those very connections for cases he or she believes warrant such efforts on the grounds of technical excellence and market potential. Similar relational combinations exist between the different categories.

Now that we are armed with the types of risks that challenge attempts to conquer new technology in general, and advanced materials in particular, we can begin to understand the different levels of performance between countries and regions. Specifically, we need to now ask in what ways have the United States and European Union differed in the means by which they have perceived and managed these various risks? Why and in what ways has the United States outperformed Europe in wrestling with and taming these potential (and actual) barriers to progress? In tackling these questions, the next few chapters explore the different phases of a new technology's life cycle, from birth to development and market placement. We begin then with the creation process itself, that is, research and development (R&D) in the United States and Europe.

REFERENCES

1. Moskowitz, S. L. (2002). *Critical Advanced Materials Report: Final Draft,* pp. 30–34 (Prepared for: Virginia Center for Innovative Technology, Herndon, VA).
2. Stuart, C. (2003), "Nanotubes are Bidding for Star Billing on Big Screens," *Small Times,* September 12; Brauer, S. (2002), "Online Exclusive: Emerging Opportunities for Carbon Nanotubes," *Ceramic Industry Online* (www.ceramicindustry.com), January 1; Holister, P., Harper, T. E., and Vas, C. R. (2003), *White Paper: Nanotubes,* CMO Cientifica (Las Rozas, Spain), pp. 1–13.
3. Moskowitz, S. L. (2002). *Critical Advanced Materials Report: Final Draft,* pp. 10–16 (Prepared for: Virginia Center for Innovative Technology, Herndon, VA).
4. Design Engineering (2003), "Nanoceramics on the Rise," Home Section, April 24; Ra-

jan, M. (2001), "US Piezoelectric Market Continues to Boom," Business Communications Company, Press Release, March 28 (Norwalk, Connecticut); Moskowitz, S. L. (2002). *Critical Advanced Materials Report: Final Draft,* pp. 24–27 (Prepared for: Virginia Center for Innovative Technology, Herndon, VA). Design Engineering (2003), "Nanoceramics on the Rise," Home Section, April 24.

5. Moore, S. K. (2003), "Just One Word—Plastics," Special R&D Report, IEEE Spectrum On-Line (www.spectrum.ieee.org), September 9; Klauk, H. (2000), "Molecular Electronics on the Cheap," *Physics World,* January, pp. 18–19; Leeuw, D. (1999), "Plastic Electronics," *Physics World,* March, pp: 31–34; Goho, A. (2003), "Plastic Chips: New Materials Boost Organic Electronics," *Science News,* Vol. 164, No. 9, p. 133; Forrest, S., Burrows, P., and Thompson, M. (2000), "The Dawn of Organic Electronics," *IEEE Spectrum On-Line* (spectrum.ieee.org), Vol. 37, No. 8 (August); Teltech Resource Network Corporation (1999), "Organic Polymer Electronics: Phase 1," Teltech Publication, Minneapolis, MN; Boulton, C. (2003), "Researchers Developing Plastic Memory Technology," *Internetnews.com,* November 14.
6. Moskowitz, S. L. (2002). *Critical Advanced Materials Report: Final Draft,* pp. 5–34 (Prepared for: Virginia Center for Innovative Technology, Herndon, VA).
7. Nanoinvestor News (2000–2005) (www.nanoinvestorNews.com), *Facts and Figures;* Moskowitz, S. L. (2002). *Critical Advanced Materials Report: Final Draft,* pp. 34–36 (Prepared for: Virginia Center for Innovative Technology, Herndon, VA).
8. "Talent Shortage Expected at Nanotech Firms" (2002). Nanoelectronicsplanet.com, January 14; Moskowitz, S. L. (2002). *Critical Advanced Materials Report: Final Draft,* pp. 25–27 (Prepared for: Virginia Center for Innovative Technology, Herndon, VA).
9. Table is constructed from information obtained in a number of sources, including: Huang, Z., Chen, H., Yip, A., Ng., G., Guo, F., Chen, Z. K., and Roco, M. C. (2003), "Longitudinal Patent Analysis for Nanoscale Science and Engineering: Country, Institution and Technology Field," *Journal of Nanoparticle Research,* Vol. 5, No. 3–4, pp. 3–17; Moskowitz, S. L. (2002). *Critical Advanced Materials Report: Final Draft,* pp. 5–27 (Prepared for: Virginia Center for Innovative Technology, Herndon, VA).
10. Moore, S. K. (2003), "Just One Word—Plastics," Special R&D Report, IEEE Spectrum On-Line (www.spectrum.ieee.org), September 9; Klauk, H. (2000), "Molecular Electronics on the Cheap," *Physics World,* January, pp. 18–19; Leeuw, D. (1999), "Plastic Electronics," *Physics World,* March, pp: 31–34; Goho, A. (2003), "Plastic Chips: New Materials Boost Organic Electronics," *Science News,* Vol. 164, No. 9, p. 133; Forrest, S., Burrows, P., and Thompson, M. (2000), "The Dawn of Organic Electronics," *IEEE Spectrum On-Line* (spectrum.ieee.org), Vol. 37, No. 8 (August); Teltech Resource Network Corporation (1999), "Organic Polymer Electronics: Phase 1," Teltech Publication, Minneapolis, MN; Boulton, C. (2003), "Researchers Developing Plastic Memory Technology," *Internetnews.com,* November 14.
11. MTI Corporation Website (www.mticrystal.com) (2003), "Online Shopping Price List"; Moskowitz, S. L. (2002). *Critical Advanced Materials Report: Final Draft,* pp. 16–24 (Prepared for: Virginia Center for Innovative Technology, Herndon, VA).
12. IP lawyers and analysts that the author has interviewed agree with this range (as of 2006). The figures also agree with the author's own experience as a consultant and expert witness on IP matters for law firms in the areas of biotech, electronics, and chemicals.
13. The table is constructed from a wide range of sources. These include Moskowitz, S. L. (2002). *Critical Advanced Materials Report: Final Draft,* pp. 15–28 (Prepared for: Vir-

ginia Center for Innovative Technology, Herndon, VA); *NanoInvestor News, Company Profile Archives.* It also incorporates recent articles on new material development, interviews with engineers and executives in the advanced materials field, and firsthand experience by the author on cases (2000–2006) involving technology assessment and impact involving the biotech, electronics, and synthetic advanced-materials fields.

14. Table is a synthesis of a number of sources, including Moskowitz, S. L. (2002). *Critical Advanced Materials Report: Final Draft,* pp. 4–25 (Prepared for: Virginia Center for Innovative Technology, Herndon, VA); Spitz, P. H. (1989), *Petrochemicals: The Rise of an Industry,* New York: Wiley. See also sources listed in Chapter 2, References 1–13 and *Nanoinvestor* News (www. nanoinvestornews.com) (2000–2005), *Company Profile Archive.*
15. Aboody, D. and Lev, B. (2001), *R&D Productivity in the Chemical Industry,* CCR Study (Council for Chemical Research, Washington, D.C.), p. 3.

Part Three

Creation: Research and Development

Chapter 5

Research and Development I: The American Context

Research and development (R&D) plays an essential part in the creation of advanced technology in general, and new materials systems in particular. Accordingly, R&D is the very source of industrial competitiveness and economic growth. It is generally understood that a significant portion of America's economic growth from 1945 to 2000 is either directly, or indirectly, related to total R&D investment. Research and development to various degrees occurs in most nations of the world and is a critical element of economic policy internationally.

Industrial research and development embodies the idea that science creation and its application to industry should be organized and directed like other corporate functions. Firms' decisions about the magnitude and nature of their R&D activity performance is guided by consideration of "returns"—economic, financial, public relations, and so on. In general, firms invest in R&D if there is a high probability of success and the expected returns meet a company's internal rate of return, and exceed those of other, viable investment options (such as acquisition of new plant, equipment, and machinery; advertising; and speculative asset purchase). Indeed, corporate histories show in detail how the day-by-day decisions, as well as long-term planning, of R&D departments hinge on revenue and profit margins, even more than on the department's scientific and technical capability.

Europe first developed the concept of the corporate R&D department during the last quarter of the nineteenth century, during the time of the creation of what were then the new materials. Germany's great chemical firms, the advanced materials companies of the day, institutionalized corporate research and development, which helped guide the companies' early success in their chemical empire, rooted in coal-based dyes and synthetics.

But it was in the United States that corporate research and development realized its full potential, especially as an engine of American mass production. As the twentieth century advanced, R&D became deeply embedded within America's growing corporate organizations and integrated with such other corporate functions as procurement, transportation, production, and marketing. Coordination of R&D activity with these other functions, realized through organizational innovations specific to American industry in the years leading up to World War I, was mandatory for the large-scale processing of raw materials, intermediates, and components into final products turned out in great volume and at low cost. The R&D department stood at the center of this system as it created the critical product and process technologies around which the mass production system evolved.

General Electric and Ford Motor Company are generally credited with nurturing the first major corporate R&D departments in the United States. But the advanced-materials firms of the time also organized important R&D activity that had enormous influence over industrial science and technology. These organizations included companies like Standard Oil of New Jersey (EXXON), the Aluminum Corporation of America (Alcoa), DuPont, Dow Chemical, Corning, Shell Chemical, and other metal, oil, and chemical enterprises.

More than any other field, progress in advanced materials historically sustained research and development programs at both the public and private levels. The first wave of new materials created in the 1930s through the 1950s emanated from early R&D departments that took root in the United States and Europe. This creative activity brought to market such important twentieth-century materials as nylon, synthetic rubber, and polyethylene.

Today, advanced materials are front and center in international R&D. Lux Capital Group LLC estimates that about $3 billion was invested worldwide on nanotech R&D in 2004, or 10% of all R&D investment. The bulk of this investment, between 75%–85%, either directly or indirectly goes towards research in advanced materials. Table 5.1 shows the steady growth in global spending in advanced-materials research.

The United States continues to lead the world in total R&D spending. Nevertheless, the European countries do not lag far behind and, in fact, have been catching up to the United States in expenditures on materials R&D. Beyond the question of amounts of money spent is the more relevant issue of effectiveness of R&D programs. Even though Europe began so promisingly in studying the science behind important advanced materials, the pattern occurs time and again of the United States taking up the actual development of the technology and making it its own. The following case studies involving four important advanced-materials groups illustrate the point.

OPPORTUNITIES TAKEN AND OPPORTUNITIES MISSED: CASE STUDIES

Advanced Polymers

The scientific basis of advanced polymers was developed in Europe. In the mid-nineteenth century, English, French, and German chemists pioneered the structural understanding of macromolecules and the first laboratory synthesis of important polymers. In Germany, for example, Hans von Pechmann first created the polymer polymethylene, a very close cousin to polyethylene. This achievement was of enormous theoretical interest, especially in regard to the makeup of the molecular units. By the 1920s, the Swedish chemist Theo Svedberg won his Nobel Prize for his ex-

Table 5.1. Global advanced-material R&D spending ($ million)

	1990	2000	2010	2020
United States	150	241	490	675
European Union	45	164	368	580
Japan	20	151	237	477
World	400	1000	1350	2200

Source: [1].

perimental work identifying polymers as "super" structures with very high molecular weights. The German Hermann Staudinger advanced Svedberg's work with his studies of rubber molecules and the first comprehensive theory of macromolecules and their formation. Around the same time, Herman Mar, another in a line of creative German scientists, used the then new technique of X-ray crystallography to further refine the theory of large molecular structures.

By the end of World War II, Europeans achieved a scientific understanding of a new generation of high-performance "linear" polymers. In the 1950s, the Italian chemist Gullo Natta and the German Carl Ziegler independently discovered a new class of catalysts that became the key to the high-density linear polyethylenes and polypropylenes that could be synthesized at low pressures. This material led the way to the advanced polymer and composite materials of the 1960s and 1970s.

Despite this impressive scientific heritage, the Europeans did not channel their theoretical and experimental achievements into commercial leadership, thus revealing a sharp divide between fundamental research and practical development. Table 5.2 traces the major achievements in polymer technology from the turn of the last century to the turn of the twenty-first century. It identifies the major product (and process) technologies, the inventor (whether individual or company), the year of commercialization, and the innovating country. (In the cases when two countries worked together on a technology, or when they worked separately but introduced it at about the same time, these are listed together.)

A number of trends can be seen from the table. In the first place, by the time of World War I, the individual inventor clearly gave way to the corporation as the source of innovation. This typically occurred in most industries as research and development grew ever more complex and expensive to carry out by the individual "genius." At this point, the R&D department evolved as a department within the large corporate structure.

Then too, despite the scientific leadership of Europe in the field, we notice the predominance of the United States as source of polymer innovation through the entire period. Certainly, Western Europe had its hand in a joint or coincident role in some important developments (polyesters and polyethylene), and in some cases moved ahead of the United States technologically. But, overall, the United States clearly controlled the pace and direction of invention and commercialization. By the 1950s, the United States dominated Europe as innovator in virtually all areas of advanced-polymer materials.

Table 5.2. Major advanced-polymer technologies 1900–1999

Technology	Company or inventor	Year	Country
Casein plastics	Adolph Spitteler	1897	Germany
Bakelite	Leo H. Bakeland	1909	United States
Urea plastics	British Cyanides Co.	1923	Great Britain
Polyvinyl plastics	Union Carbide; B.F. Goodrich	1925	United States
Ethylene	Union Carbide	1930	United States
Acrylic	Rohm & Haas	1931	Germany
Melamine resins	American Cyanimid, CIBA	1933	United States, Germany, Great Britain
Polystyrene	Dow Chemical, I.G. Farben	1935	United States, Germany
Nylon	DuPont	1938	United States
High-pressure polyethylene	DuPont, ICI	1939	United States, Great Britain
Epoxy resins	IG Farben	1939	Germany
Polyurethane	IG Farben	1940	Germany
Polyesters	Calico Printer's Association	1941	Great Britain
Fluid catalytic cracking*	Standard Oil New Jersey (now Exxon)	1942	United States
BUNA Rubber	Standard Oil New Jersey (Exxon), IG Farben	1942	United States, Germany
Polyethylene	DuPont, ICI	1944	United States, Great Britain
Polyester			United States, Great Britain
Ziegler catalyst, first commercial application	B.F. Goodrich, Gulf Oil	1954	United States
Polytentrafluoro-ethylene (PTFE), "Teflon"	DuPont	1955	United States
Phillips polyethylene process	Phillips Petroleum	1956	United States
High-density polyethylene	Hoechst, W.R. Grace, Hercules, Phillips	1956	United States, Germany
Crystalline poly-propylene	Hercules, Montecantini, Hoechst	1957	United States, Germany, Italy
Stereo-specific rubber	Goodrich-Gulf, Phillips, Shell	1959	United States
Silicone plastics	Dow Chemical	1959	United States
Acrylonitrile	Sohio	1960	United States
"Dabco" catalyst process for urethane	Houdry Corp.	1960	
Polycarbonate plastics	General Electric	1975	United States
Kevlar	DuPont	1978	United States
UNIPOL process of linear low-density	Union Carbide, Shell	1980s	United States

Table 5.2. *Continued*

Technology	Company or inventor	Year	Country
polyethylene and polypropylene			
"Cyclar" process for production of aromatics from LPG	UOP	1999	United States

Source: [2].

Organic Polymer Electronics (OPEs) [3]

The field of organic polymer electronics gives us a telling example of this gap in fundamental research and practical development between the United States and Europe. As discussed in Chapter 2, these materials are most important members of the advanced-material group.

The signal scientific work in OPEs took place in Europe. In the 1980s, both England and France independently conducted fundamental and pioneering scientific inquiries tracing the relationship of intermolecular bond structure and semiconductor properties in synthetic organics. From this work, European scientists developed laboratory-scale OPE devices, such as the first small-molecule, organic thin-film transistors. Although these devices had little commercial use, they aided further scientific investigation in the field.

In England, Cambridge University's Cavendish Laboratory, a center of scientific inquiry since the nineteenth century, extended the theoretical envelope by pursuing an improved conceptual understanding of carrier mobility within semiconductors in general, and OPEs in particular.

This work in England and France led to the publication of a flurry of important scientific articles. The United States certainly contributed to this science, but was hardly at the forefront of the theoretical work. Rather, it tapped into this growing science, adapted it, and added to the mix a practical understanding of the way organic polymer semiconductors worked (or could work) in an industrial setting. Development based on engineering design, not pure science, fueled this effort. In the United States, the locus of activity shifted from the university science departments to the company and its engineering laboratories.

In the late 1980s and early 1990s, Eastman Kodak Co. (Rochester, NY) conducted innovative engineering design, as opposed to theoretical, studies in this field and, as a result, produced the first efficient electron emission pattern from a two-layer organic structure resembling a p–n junction, thus leading the way to working organic light emitting diode (OLED) technology. This technology, the first to demonstrate acceptable light emission at low voltages, attracted the interest of potential display manufacturers, particularly in the United States and Japan.

Following up on this engineering work, Xerox examined new types of semiconductor organic polymers to serve as flexible plastic circuits for portable, "flat" television screens and monitors made from a single sheet of plastic. In contrast to the Europeans, Xerox did not study these materials just to understand their physical structure; rather, they wanted to discover the limitations of existing materials in order to develop "design rules" to get around these barriers to commercialization. Xerox and other companies evaluated these materials in simple electronic display devices with the cooperation of leading electronics firms around the world.

Since that time, attention has turned to development of new and more effective organic polymer semiconductors and on efficient ways in which to manufacture the technology. This work was again centered in the United States and on engineering design applications. In the late 1990s, Lucent Technology, in cooperation with Northwestern University, devised a new class of organic semiconductor materials—rod shaped thiopene (polythienylenevinylene and polyimide compounds) organic molecules—that is expected to accelerate commercial application of OPEs, especially in such disposable plastic electronic devices as smart cards, electronic tags for tracking inventory, and chemical sensors. In addition to Lucent, such U.S. firms as Phillips Electronics, Sarnoff Corp., and Hewlet Packard improved manufacturing processes for flexible, all-polymer integrated circuits using such inexpensive techniques as spin-and-dip coating, casting, vapor deposition, and printing techniques that do not require costly vacuum technology.

Nanocarbon Materials: Metal Fullerenes and Nanotubes [4]

Also in the 1980s, advanced-material research, for the first time, extended beyond the field of polymers into other promising areas, including nanocarbon materials. As noted earlier, nanocarbon materials comprise open molecular structures defined by the number and specific arrangement of carbon atoms and are capable of a range of commercial possibilities. Metal fullerenes are a particular type of nanocarbon material; they contain one or more metal atoms surrounded by an arrangement of carbon atoms. Nanotubes are nanocarbon fullerenes that consist of a closed tubular array of carbon atoms.

As with the polymer group, Europe produced some of the earliest and most fundamental scientific studies in this field. As early as the 1860s, the German chemist August Kekule first proposed the circular structure of organic (carbon-based) molecules. Subsequent work over the next century strove to determine the basic structure of important organic materials. The Europeans continued to excel here. The work of scientists such as Herman Staudinger during the first half of the twentieth century extended the achievements of Kekule to macromolecules. By the 1970s, Europe again led the way in the scientific inquiry of the new materials. Fundamental work in carbon nanomaterials was conducted by Prof. Morinobu Endo, who discovered carbon nanotubes while a researcher (from Japan) at France's University of Orleans. He observed straight, hollow tubes of carbon produced by a gas-phase process. A few years later, extending Endo's work, scientists at the University of Montpelier succeeded in a laboratory-scale process for synthesizing nanotubes for research purposes.

These early scientific accomplishments notwithstanding, it was again in the United States that the technology, in contrast to the science, flourished. Borrowing and adapting whatever parts of the science proved useful to application and employing the methods or engineering design, IBM, in 1993, isolated for the first time the industrially important single-walled nanotubes, soon to be known as "buckeyballs," in honor of Buckminster Fuller and his famous geometric design innovations.

Commercialization of nanotubes took the next important step in 1996 when Richard Smalley and coworkers at Rice University synthesized carbon nanotubes through the use of pulsed laser vaporization of a carbon target, a furnace fired to 1200°C, and a cobalt–nickel catalyst to accelerate nanotube formation. This pulsing induced a more uniform vaporization and better control of the growth conditions than previously possible. This technology has had the greatest impact in the field, due to its efficiency and quality of product.

Currently, the United States continues to extend the technological envelope in the field, even as Europe continues to pursue the science. Most of the important companies that have been involved with commercializing nanotubes and fullerenes, including Carbon Nanotechnologies, Eikos, Luna Nanoworks, Nano C, and Southwest Technologies, are based in the United States.

Nanocrystals and Quantum Dots [5]

Not surprisingly, the scientific foundations of nanocrystals originated in England in the nineteenth century. In the1850s, Michael Faraday examined the structure of microstructured crystalline systems. His work entailed an investigation of the electrooptical properties of crystalline materials. Further inroads into the science had to wait until the development of quantum theory after World War I and the clearer understanding it provided of the electronic structure of crystal systems. German, French, and British scientists actively pursued the field. Single-electron quantization (tunneling) within crystal structures pushed the scientific frontier after World War II.

Then, in early 1980s, Russian scientists undertook important work in the science of nanocrystals and quantum dots. Alexander Efros and A. I. Ekimov of the Yoffe Institute in St. Petersburg (then Leningrad) in the former Soviet Union explored the so-called "quantum confinement" effect. A decade later, German scientist Horst Weller wrote the first-ever review of the state of the science of semiconductor nanocrystals that synthesized past and current theories in the field.

But it was in the United States that commercialization progressed. Researchers at MIT and Los Alamos National Laboratory showed conclusively that quantum dots behave as semiconductors and can provide efficient emission of laser light necessary for creating such optical-electronic technology as tunable lasers, optical amplifiers, and LEDs. With the work done by Bell Labs in the late 1980s, commercial development of nanocrystals proceeded. Here we see the interest in engineering design as opposed to scientific or engineering theory. Bell scientists (Louis Brus, Moungi Bawendi, and Paul Alivisatos) made seminal experiments leading to potential applications. They experimented with nanocrystal semiconductor materials and observed solutions of strikingly different colors made from the same substance. In a series of engineering design studies, they borrowed aspects of the quantum confinement effect to develop commercially relevant laws that correlated the size and color of nanocrystals. These investigations established the engineering principal that the physical properties of substances change when the crystal structures enter the nanosize range. These results led to development of nanocrystal transistors and other applications.

The IBM engineers who worked on nanocrystals then left the company to teach at such universities as the University of California at Berkeley and MIT, and to further investigate practical applications of nanocrystals. From this work, the material for the first time could be made soluble in water, an important step leading to further applications. Through empirical research, American engineers then discovered that adding a passivating inorganic "shell" around nanocrystals in solution, and then shining blue light on them, caused the nanocrystals, or quantum dots, to light up brightly. This led to the formation of Quantum Dot Corp., which licensed these discoveries for medical diagnostic applications. Soon thereafter, Nanosys signed an exclusive licensing agreement for use of the materials in light emitting diodes (LEDs) for lightweight computer displays. Currently, only American companies, including Quantum Dot Corporation, Nanosys, Applied Nanoworks, and Evident Technologies, are recognized as leaders in quantum dot technology.

THE UNITED STATES, R&D, AND ADVANCED MATERIALS

Why has the United States been so successful in taking what had been brilliant scientific beginnings from elsewhere and creating living and breathing technologies? We are not here talking about adaptation, although this in itself is no mean feat, but actually turning raw material into fully realized systems of economic growth. If Europe has been, and continues to be, the epicenter of scientific progress, the United States has been and remains the global leader in practical engineering design leading to economically useful products. This distinction is clearly reflected in the focus of R&D spending and in patent trends.

What Type of R&D?

Beginning in the 1980s, the United States has reasserted itself as the center of the world's research and development activity, especially in that part of R&D that looks to transport new ideas into the commercial sphere. A look at R&D expenditure trends is instructive. Between 1989 and 1999, total US real spending on all R&D rose more than 36% (adjusted for inflation and 1992 constant dollars). The proportion of Gross Domestic Product devoted to R&D in the United States during the second half of the 1990s increased from 2.6% to nearly 2.8%, a rate higher than any year since 1967. Not surprisingly, Japan remains one of the most important countries in carrying out advanced R&D today. During the late 1990s, R&D spending in Japan as a part of GDP grew 1.8% annually. By 2003, Japan invested 3% of its total GDP to research and development activities.

A more important measure of R&D effectiveness is the portion of this activity accounted for by industry. The United States remains the leading performer of corporate R&D worldwide by a clear margin. Private industry's share of R&D funding rose from 50% in 1987 to more than 68% in 2000. Industry in the United States accounts for roughly 45% of the industrial world's investment in R&D. Industrial R&D expenditures in the United States are greater than all industrial sectors of the European Union combined, and twice the industrial R&D performed in Japan.

Traditionally, the type of R&D focused on by the United States differs from that carried out in Europe. U.S. industry spends less of its resources on basic or theoretical scientific inquiry and the greater portion on applied and engineering development. Within the United States, over 75% of money spent on R&D within industrial and government laboratories goes to such type of engineering-related work. Within Europe, this figure is under 50% with the rest devoted, in some form or another, to basic scientific research.

Where is R&D Located? The Decentralization of Innovation

Not only is American R&D biased less toward fundamental research and more to development and commercialization, but, as discussed in the last chapter, over the last half century, a major shift has occurred within the United States from centralized to decentralized, whereby, the locus of R&D relocated from the federal government and large corporate subcontractors to localities working with a larger number of smaller, highly innovative firms.

Table 5.3 reflects this trend. Summarizing the results of the important advanced-material innovations, it shows the percentage of these technologies that relied (or are anticipated to rely) in a significant way either directly or indirectly on federal versus local support (e.g., funding, R&D support, tax incentives, etc.). From the table, we note the growing role of federal (or central) governments from 1940 through the 1960s. Contraction of federal funding began in the 1970s. With the end of the Cold War in 1989, federal funding of R&D continued to decline at a rapid rate. In contrast, state and local government has steadily increased its influence over the innovation process. From 1970 until 2000, local funding of R&D grew from 15% to 45% [7]. This is not true internationally, where, as we have noted, the central governments maintained robust support of basic research.

Table 5.3. Distribution of government input into major advanced-material technology, 1940–2030, United States (% of innovations)

	Federal (%)	Local (%)
1940	45	10
1950	55	10
1960	60	15
1970	65	15
1980	55	30
1990	45	40
2000	30	45
2010	25	55
2020	20	60
2030	15	65

Source: [6].

The Decentralized Model: A Question of Economic Growth

The paradigm of government–industry R&D alliances has clearly changed since World War II, from R&D targeting military and space-related outcomes to a focus on economic and social benefits. But the ability of federal government to achieve these goals, especially at the state and local level, is quite limited. State and local jurisdictions have far greater incentive to create economic growth, jobs, and fiscal flexibility. In some cases, federal, state, and local governments act in consort. For example, the National Science Foundation attempts to apply national monies to specific regions with high technology potential. But because the situations and requirements differ widely between the federal and state and local governments, states, counties, and cities have increasingly taken it upon themselves to identify and support local technology centers. Within the United States, the number of states and cities participating in technology growth initiatives has been increasing at a rapid rate. Just a decade ago, less than 15% of U.S. states pursued advanced-technology initiatives. By 2004, this figure had grown to over 50%. By 2010, it is expected that at least 80% of all states will have such initiatives in place.

Localities view new technology and the businesses and industrial clusters surrounding them as vital to their region's economic growth, with job creation the most sought after benefit. The pressures of globalization, which state and local politicians think will lead foreign companies and technologies to flood into their neighborhoods, forced elected authorities and the businesses communities which they serve to take defensive competitive actions, which is reflected in the growing percentage of government funding dedicated to R&D at the state and local levels (Table 5.4).

In supporting "localized" research and development initiatives within the advanced technology fields, especially new materials, the states hope to revitalize once-thriving areas. By 2005, most states either had put into place some form of nanotechnology initiative, of which advanced materials is such an important part, or were considering doing so. Thus we find that Oklahoma established its own Nanotechnology Initiative to create statewide awareness and attract funds for new industry in the state. In some cases, states actually support individual companies who work within promising areas. Ohio supports particular companies in such areas as power devices and fuel cells. In Illinois, the state and federal government work jointly to develop new-technology activity. Together, they provided $17 million in 2003 toward the construction of a Center for Nanoscale Materials at Argonne National Laboratory.

In Mississippi, we observe the collaboration between state, county, and city governments in the support of economic growth through advanced-materials R&D. Recently, the city of Hattiesburg, Mississippi struck a deal between Hybrid Plastics (a California polymer company), the University of Southern Mississippi (USM, an important polymer center), and the county's Area Development Partnership (ADP) to bring the company to USM's Hattiesburg campus. Hybrid Plastics is building a 1500 sq ft laboratory within USM's Polymer Science Building. This agreement allows USM and Hybrid Plastics to collaborate in the research and development of polymers and derivative nanocomposites. The ADP and the Forrest County Industrial Park are supporting construction of a 26,000 sq. foot manufacturing facility within the city out of funds borrowed from the Mississippi Development Authority. From this beginning, the parties are establishing a Nanotechnology Center of Excellence at SMU that will concentrate on applying the technology to improve agricultural products, electronic polymers, coatings, paints, and composites.

Rivalry between states for superiority in advanced-materials R&D continues to grow as the implications for economic development and job growth become clear. New York and Texas, in particular, have been competing for prominence. Texas is looking to revitalize its technology corridor in and around the Austin area, and New York is attempting, with some real success, to spur industrial growth in economically depressed upper New York State. Both states follow a similar model for growth: entice leading consumers of advanced materials into the region, link these companies with nearby universities, and create an expanding R&D and industrial network of complementary industries and firms. The center of interest in this partic-

Table 5.4. Government support of advanced-materials R&D

	1950	1960	1970	1980	1990	2000	2010	2020
Federal support of R&D (% of total government money supporting advanced-material R&D)	80	85	60	50	45	40	35	30
State and local support (% of total government money supporting advanced-material R&D)	20	15	18	30	55	60	65	70

Source: [8].

ular rivalry is the international semiconductor chip consortium, Sematech. Both states wanted to use Sematech as the seed around which a large, diverse, and growing R&D community would grow. The advantages of Texas include an already established high-tech community formed from capturing a chunk of the semiconductor industry in the 1970s and 1980s, and robust university life centered around the University of Texas and Texas A&M University.

But New York State, particularly in the Albany area, has been gaining ground on Texas in terms of advanced-materials R&D. It is currently one of the most active centers in this arena. Political direction is proving a critical component of the program. During his tenure, Governor Pataki spearheaded the effort and helped direct money and technical resources to the area. The Governor of New York has, in general, significant financial and political advantages over his counterparts in other states. He oversees one of the largest state budgets in the United States and wields greater administrative and financial power than the Governor of Texas. Moreover, New York's Governor is not required to balance the budget and so can offer R&D companies greater financial incentives than other states. The Governor's office is also empowered to aggressively recruit high-technology projects.

Under Governor's Pataki's watch, New York State moved fairly rapidly in forming an advanced research center in Albany with over 1000 researchers, including a number from Sematech. From this advantageous point, New York attracts investment from both within and outside the state, a critical component in continuing to entice high-technology companies to utilize the benefits of the research center, including semiconductor and new-materials companies. Just as the United States appears to dominate new-materials R&D globally, so New York State has carefully positioned itself to be a central player in twenty-first century R&D nationally.

What Are the Fruits of R&D? Patents

Although the Federal government's role as source of practical R&D has been reduced within the United States compared to other countries, nevertheless, it has served in a way that actually made it easier for universities and local actors to join hands in creating technologically fruitful businesses. The fact that the R&D system must operate "close to the ground," as it were, means that there should be an intimate and vibrant linkage between R&D, patents, and the marketplace. This connection is clearly seen through the growing efficiencies we note in the process by which advanced material as well as other types of patents are transferred form place of origin, whether government laboratory, university, or industry, to the commercial world, and, in particular, the innovative start-up firm. These efficiencies emerged through regulation and the rise of the innovative and ambitious small- and medium-sized enterprise ensconced within local cluster groups.

Once the university creates useful knowledge, patenting (and licensing of patents) is the important step that helps create a commercial enterprise that makes use of that knowledge. In examining patent data, we note that there are critical indicators showing the close relationship between U.S. research and development and the commercial world. In the first place, the United States, as we have seen, is more productive than Europe in turning out advanced-material patents. Then too, an increasing "closeness" between patents and the marketplace becomes evident in the narrowing length of time taking place to go through the various stages for applying for, being granted, and finally "working" a given advanced-materials patent.

The patent structure of the United States is vitally important as the medium by which the results of advanced-materials research and development becomes commercial reality as start-up companies and spin-offs. Licensing is the primary mechanism. Whereas European patent policy places barriers between R&D and the market, American policy works to ease commercially relevant research into the market. This harmonization of research, patents, and markets reached an important benchmark in the early 1980s when the United States passed the Federal Courts Improvement Act (1982) that significantly upgraded effectiveness and efficiency of the patenting process. This piece of legislation, in conjunction with the Bayh–Dole Act, extended patenting privileges to inventors in universities and government laboratories, encouraged the formation of cooperative arrangements between laboratories and private firms, and permitted universities to automatically retain title to patents derived from certain types of R&D, thus removing the need to get a waiver from the funding agency in order to exploit patent rights.*

Studies show that both universities and laboratories increased their patenting between the late 1980s and the mid-1990s. If not sufficient by themselves, these legislative changes were at least necessary in expanding the licensing of advanced-material patents to start-up companies and spin-offs, in large part because they gave greater patenting freedom to universities as well as security to licensees.

Table 5.5 illustrates the point. It shows that the licensing by U.S. start-up and spin-off firms in advanced materials, many working in concert with state and local funding agencies and technology centers, began to accelerate in the 1980s.

The powerful combination of government regulation of the patent system, useful research by university engineering (as opposed to science) departments, and rise of the creative small start-up in conjunction with local agents of growth, has led to a growing "closeness" between patents and the marketplace. Table 5.6 shows historical trends in the number of advanced-material patents granted within the United States and the percentage of these that are actually commercialized. The surge in

*At this time, the government created a new court to review patent litigation decisions and improve the chances for success in court for patentees. This was required because the greater complexity of technology, such as advanced materials, proved an incentive for imitators to try their luck in the courts with challenges to earlier and fundamentally valid patents. In effect, the Act standardized patent law across the country and eliminated incentives for frivolous patent challenges. This, in turn, reduced the chances for success for challenges by imitators. Before 1980, a district court finding that a patent was valid and infringed was upheld on appeal 62% of the time; by 1990, this percentage rose to 90%. Conversely, before 1980, appeals courts overturned only 12% of district court findings of patent invalidity or noninfringement; that percentage rose to 28% by 1990. With the risk of failure of their suits greatly increased, newcomers have become less interested in challenging existing patents [9].

Table 5.5. Licensing trends in advanced materials, United States

	Percentage of all U.S. advanced material firms that licensed from universities and government
1960	3%
1970	6%
1980	8%
1990	18%
2000	26%
2004	32%

Source: [10].

Table 5.6. Patent trends in advanced materials: number granted and percentage of patents commercialized, United States

	1930	1950	1960	1970	1980	1990	2000	2010
Number of patents	650	760	575	430	570	675	1470	2455
Percentage of patents commercialized	5	8	6	7	10	12	15	20

Source: [11].

both of these is a strong indicator of the growing "marketability" of American advanced-materials research.

REFERENCES

1. Sources for construction of the table include: Commission of the European Communities (2004), "Towards a European Strategy for Nanotechnology," in *Communications from the Commission* (Brussels, Belgium), pp. 24–25; Dunn, S. and Whatmore, R. W. (2002), "Nanotechnology Advances in Europe," *Working Paper, Directorate-General for Research* (European Parliament: Luxembourg), pp. 6–10.
2. Spitz, P. H. (1989), *Petrochemicals: The Rise of an Industry* (New York: Wiley): 5–165; Aftalion, F. (2001), *A History of the International Chemical Industry* (Philadelphia: Chemical Heritage Press): 6–150.
3. Moore, S. K. (2003), "Just One Word—Plastics," Special R&D Report, IEEE Spectrum On-Line (www.spectrum.ieee.org), September 9; Klauk, H. (2000), "Molecular Electronics on the Cheap," *Physics World,* January, pp. 18–19; Leeuw, D. (1999), "Plastic Electronics," *Physics World,* March, pp: 31–34; Goho, A. (2003), "Plastic Chips: New Materials Boost Organic Electronics," *Science News,* Vol. 164, No. 9, p. 133; Forrest, S., Burrows, P., and Thompson, M. (2000), "The Dawn of Organic Electronics," *IEEE Spec-*

trum On-Line (spectrum.ieee.org), Vol. 37, No. 8 (August); Teltech Resource Network Corporation (1999), "Organic Polymer Electronics: Phase 1," Teltech Publication, Minneapolis, MN; Boulton, C. (2003), "Rcsearchers Developing Plastic Memory Technology," *Internetnews.com,* November 14.

4. Stuart, C. (2003), "Nanotubes are Bidding for Star Billing on Big Screens," *Small Times,* September 12; Brauer, S. (2002), "Online Exclusive: Emerging Opportunities for Carbon Nanotubes," *Ceramic Industry Online* (www.ceramicindustry.com), January 1; Holister, P., Harper, T. E., and Vas, C. R. (2003), *White Paper: Nanotubes,* CMO Cientifica (Las Rozas, Spain), pp. 1–13; Moskowitz, S. L. (2002). *Critical Advanced Materials Report: Final Draft,* pp. 16–24 (Prepared for: Virginia Center for Innovative Technology, Herndon, VA).
5. Ouellette, J. "Quantum Dots for Sale," *The Industrial Physicist Online* (www.aip.org), Vol. 9, Issue 1, pp. 14–17.
6. Sources for construction of the table include 3i (2002) *Nanotechnology—Size Matters: Building a Successful Nanotechnology Company;* Moskowitz, S. L. (2002). *Critical Advanced Materials Report: Final Draft,* pp. 5–30 (Prepared for: Virginia Center for Innovative Technology, Herndon, VA).
7. MIT (2002), "Charles M. Vest, President, MIT Economic Forum, Baylor University August 13 2002," Press Release, MIT News Office, August 23, (MIT, Cambridge, MA).
8. Table constructed from a number of sources, including this Chaper, References #s 6 & 7.
9. Huston, J. (2003), "Litigating Patent Rights in a Down Economy," *Massachusetts Lawyers Weekly,* October 13, pp. 1–3.
10. Table constructed from information provided by websites of critical advanced-materials companies in the United States. Also, *NanoInvestor News* (www.nanoinvestornews.com) (2001–2005), *Company Profile Archives.*
11. Table constructed from a number of sources. See this Chapter, References 6 and 7. Also, refer to Huang, Z., Chen, H., Yip, A., Ng., G., Guo, F., Chen, Z. K., Roco, M.C. (2003), "Longitudinal Patent Analysis for Nanoscale Science and Engineering: Country, Institution and Technology Field," *Journal of Nanoparticle Research,* Vol. 5, Nos. 3–4, pp. 5–15.

Chapter 6

Research and Development II: The European Context

Europe recognizes the importance of advanced technology to industrial competitiveness. In 1999, the European Commission concluded that "Research and technology [can] account for up to 50% of economic growth and have a strong influence on competitiveness and employment and the quality of life" [1]. At a summit conference in Lisbon in 2000, European Union officials stated that, by 2010, it wanted to become the most competitive "knowledge-based" economy. However, Europe concedes it has yet to secure an acceptable level of industrial competitiveness: "In Europe . . . the situation concerning [practical] research is worrying. . . . [This] could lead to a loss of growth and competitiveness in an increasingly global economy" [2] This problem arises in terms of employment of researchers, trade balance in high-tech products, and other indicators.

More troubling for Europe is its acknowledgement that the United States enjoys stronger growth in innovation and knowledge diffusion than the EU. A 1999 report on biotechnology clusters by the U.K. Ministry of science believes that the United States is blessed with a "can-do" mentality that contributes greatly to the country's economic success. Comments by British entrepreneurs working in the United States emphasize this factor of positive thinking—a belief that anything is possible—and an American sense that failure is not to be feared because one can learn from one's mistakes as the main reasons they decided to abandon Europe to work in the United States [3].

The output of patents in advanced materials—an approximate indicator of technology creation—is instructive. It was in fact the United States and Japan that began to publish advanced-materials patents in the 1970s, only to be followed by such EU countries as France, the United Kingdom, Switzerland, the Netherlands, and Italy. For the period 1976 to 2002, the United States and Japan led all other countries in patents. While it is true that the United Kingdom and France are among the top five patent-producing countries, they both trail the United States and Japan, whereas Germany does not even rate a position in the top ten countries. Overall, although the European Union dominates the science, it accounts for only 36% of world nanotechnology patent production (as of 2004). Moreover, the rate of patenting in the European Union continues to fall behind that of the United States, even as its scientific production surges ahead [4].

As stated in the 2003 European R&D Performance Report, "Europe . . . while being strong in terms of scientific performance, has proven to be weak in terms of patenting or the conversion of scientific knowledge into products with industrial and economic benefits" [5]. So we find in biotechnology, an important sector for advanced materials, a clearly growing trend in patents granted U.S. companies worldwide. The European Union presents a very different picture. In the late 1980s and 1990s, while Europe experienced high growth in patent *applications,* the number of patents actually granted within the European Union declined. In fact, "At the European Patent Office, where European firms could be expected to have a home advantage over U.S. firms, U.S. firms account for a larger share of biotech patent applications (51.9%) than E.U. firms (27.8%)" [6].*

Table 6.1 looks at trends in three indicators of technological activity: proportion of all global patents coming from the European Union, trade deficit with the United States in high-technology products, and the portion of all advanced materials firms based in the Western E.U. countries. The table indicates that the European Union is realistic in being concerned over its performance and its future ability to compete with the United States (and Asia).

In fact, within the high-technology sector generally, and nanotechnology and advanced materials in particular, the European Union has witnessed a decline for a number of decades. We have seen, for example, that Europe left the field of advanced polymers to the United States starting as early as the late 1950s. Overall, the European Union has a deteriorating position in terms of high-tech trade. The European Union's trade deficit in high-tech products grew from 9 billion Euros to 48 billion Euros between 1995 and 2000. This gap between the European Union and the United States in high-technology trade continues to grow, even in the face of America's overall trade deficit.

THE EUROPEAN DILEMMA IN R&D I: FUNDING PROBLEMS

A variety of forces have been acting on Europe that have severely limited the rate and extent of its technological output through R&D activity.

The European Union has been falling behind the United States in its support of R&D. In the latter half of the 1990s, the portion of Europe's gross domestic product captured by R&D activity expanded at only a modest 0.6% growth per year. In 2000, the gap between U.S. and E.U. (total) investment in R&D was 124 billion Euros, a difference that has been doubling at constant prices since 1994. This relative-

*Today, the top 14 countries with the largest number of nanotechnology patents (in the period 1976–2002) are (in order): United States, Japan, France, European Union, Taiwan, Korea, Netherlands, Switzerland, Italy, Australia, Sweden, Belgium, Finland, and Denmark [7]. The United States, experiencing a longer and more significant surge in actual patenting than the European Union, enjoys stronger growth in innovation and in knowledge diffusion. Then too, the United States has patented internationally to a far greater extent than the European Union, indicating that E.U. innovators are not as involved in international market development and knowledge diffusion as their U.S. counterparts.

Table 6.1. Indicators of technological decline of the European Union

	1980	1990	2000	2010
Portion of global nano and advanced-material patents accounted for by the EU (%)	45%	45%	35%	30%
High-technology trade deficit (billion of Euros)	2	5	48	55
Portion of total advanced materials firms that are based in the EU (%)	12%	12%	10%	7%

Source: [8].

ly poor showing on the part of the European Union reflects the fact that the European countries that have been the strongest competitors to the United States—France, Germany, and the United Kingdom—lost momentum in their R&D activity starting in the 1990s, with the European Union's other industrialized countries finding themselves falling behind these "leaders."

But the problem for Europe does not simply hinge on R&D expenditures. Even if it were to pour more funds into research and development activity, significant structural problems remain. Europe has undergone a significant transformation over the last decade. Today, the European Union is a full-fledged economic entity composed of 25 nations. It commands an area and economic region similar to that of the United States. Indeed, the European Union sees the United States, a confederation of states under a central government, as a model for its own regional integration. With a centralized governing structure and a Europe-wide parliament gaining in decision-making power, the European Union is evolving from an economic to a political union. Only with integration, Europe believes, can it hope to compete against the size and economies enjoyed by the American market. Thus, competition lies at the heart of E.U. strategy. But on a number of fronts, the EU finds itself facing an uphill battle as it strives to gain the initiative as an effective region of research and development activity.

THE EUROPEAN DILEMMA IN R&D II: STRUCTURAL PROBLEMS

Structural Problems I: The Questions of Patent Law and Resource Allocation

The European Union's patent structure is fraught with hindrances that go far in explaining, at least in part, the gap in patenting between Europe and the United States. In part, this slowdown in patenting might be explained by administrative changes taking place in the European Union. There are now two patent systems in Europe that have not yet been harmonized: the older national patent systems and the centralized European Patent System. To a greater extent than in the United States, patenting in the European Union is expensive and troubled with legal uncertainty.

This bifurcated structure causes significant problems for the patenting of E.U. technology at home and internationally. Delays occur as administrative conflicts resolve themselves between the E.U. and national systems. Uncertainty in the timing of decisions, or even that decisions made by one authority could be overturned by another at a later date, play havoc with business plans and market entrance schedules by companies. The financial and manpower resources needed for pursuing legal matters also cut seriously into R&D budgets of companies, thus further delaying and even jeopardizing future innovative efforts. This is particularly true of start-up companies, which are the important players in advanced-materials research. This cumbersome mechanism serves as a brake to industrial competitiveness since it is critical to be able to shorten development time and product cycles, to react quickly and flexibly to new opportunities and economic demands in order to capture first-mover advantages in markets.

But other problems also come into play. Europe faces a resource distribution issue. In the European Union, money for research is allocated based on the principle of *juste retour,* by which countries receive a share of research and development funding proportional to their contribution to E.U. revenue. Their share of R&D money, therefore, is not necessarily based on merit, as determined, for example, by the appropriateness and promise of a project on market impact. With bureaucrats thus setting research priorities, the European Union does not follow the most advanced or most economically promising R&D agendas. European Union policy continues to support centralized R&D by financing and otherwise advancing large, centralized industrial laboratories and "big science" projects. These consortia are led either by large university laboratories or established industrial concerns. In the former case, there is a great emphasis on basic or theoretical science, and in the latter on quick-profit, short-term projects. With the European Union's continued preference for these large-scale R&D projects, both the government and institutional banks continue to avoid any significant support of the European-based start-ups. In such an environment, the start-up firm appears as a maverick company that at best may be tolerated but certainly not actively encouraged. Through the 1990s, these firms remained the renegade "outliers" of the European R&D landscape. Consequently, relatively few high-technology start-up firms are European based. The more advanced European companies that want to expand into new technology tend to enter into collaborative R&D agreements with U.S. firms rather than with other European firms.

Structural Problems II: The Universities, Science, and innovation: The Question of Entrepreneurship

The European Union's structure is such that it fosters a shortage of university graduates with the real-world, entrepreneurial mind-set so critical for the creation of practical technology. Europe's traditional strengths, as embedded in its university curricula, remain solidly rooted within the pure sciences, rather than in industrial practice. The career path for successful academic scientists in Europe tends to keep them from applying what they have learned for lengthy periods of time. For example, after graduate and postdoctoral work, European scientists can achieve university tenure by remaining in the academy and writing what is a de facto second dissertation based mainly on theoretical pursuits. This system keeps the scientists within the academy, and at a distance from commercial work, longer than their counterparts in the United States, where there is a long-standing tradition of academic–industry cooperation during graduate and postgraduate studies. Even into the twenty-first century, European engineering departments still put greater weight on theory and peer-reviewed articles than on practice and patenting. European universities also do not encourage spin-offs from research conducted within their walls. They do not regularly place engineering students in industry settings to apply coursework in a commercial setting. This "big science" bias hinders the formation of a robust entrepreneurial spirit in graduates, who tend to prefer entering the workforce as employees of large corporations rather than as managers of start-ups.

There is growing concern over this situation within European Union's advanced-technology community. According to the "Nanotechnology in Europe" (2007) survey, sent to organizations and companies throughout the European Union, European advanced materials and nanotechnology industries believe the business world is not sufficiently understood by university professors. The survey condemns European universities as being out of touch with the complex processes of technology creation and of "scientificizing" technology. That is to say, they treat technology as if it just falls naturally as fruit from the tree of theory. As a consequence, graduates from European universities view technology simplistically as if it were just an extension of scientific principle. The result is an overemphasis on the time-consuming process to "perfect" a technology as a system in all aspects, whether essential or not to final commercialization. In light of this, it is less than surprising that, as the researchers of the study conclude:

> Fewer European scientists than U.S. scientists give careful consideration to the real prospects of successful commercialization of their discoveries and inventions. . . . Europeans tend to focus on optimizing technology rather than business development and by the time the "perfect product" is available, it is often out of date and too expensive. . . . [9]

Fundamentally, there is too much looking for overarching "technology platforms" in the same way as looking for a broad scientific theory, whereas (as in the United States) there should be more emphasis on developing technology solutions addressing specific market needs.

This weak linkage between university and industry also concerns the 1999 report on biotechnology clusters. The report tells us that the United States, more so than Europe, supports a structure allowing more leeway for technology transfer from university to industry. In the United States, there are more "arrangements [in universities] whereby researchers are allowed a significant number of days a year for consultancy and commercial activities," to keep in touch and interact with the outside world. The report points especially to the "work of the MIT Entrepreneurship

Center and the role it plays in teaching entrepreneurship to MIT engineers (with courses covering the nuts and bolts of business plans, starting and building a high-tech company, and new product and venue development). We believe that British universities can learn from such courses . . ." [10].

With these specifically European forces at work, it is not surprising to find a glut of pure scientists chasing advanced-materials science with little chance of translating their work into commercial reality. Europe faces the prospect of a shortage of personnel capable of spanning the scientific, engineering, and commercial realms—the gatekeepers—who translate technical achievement into market products. This dearth of broad-based, real-world entrepreneurs, the type that animate and ultimately transform the U.S. industrial landscape, clearly hinders European conquest of the new technologies. European participants in the nanotechnology industry understand all too well that

> . . . an entrepreneurial culture is missing among scientists in Europe . . . business support, leadership and guidance is lacking, and the recognition for successful exploitation of science is also lacking. . . . Europe has no foundation or track record for serial entrepreneurs . . . [whereas in the United States] nanotechnology companies and scientists [are entrepreneurial in that they] actively seek investors. [11]

This low level of "entrepreneurial dynamism" in Europe, so much a product of the fissure between theory and practice, and between academia and industry, manifests itself in the indices of entrepreneurial activity (the number of people who are defined as entrepreneurs per 100 adults (aged 18–64). Whereas (in 2003) the United States registered 12%, the major economies of Europe each came in between 7.7% and 7.0%.

Structural Problems III: Resource Overextension: The Question of Eastern Europe

There is also a resource overextension problem, especially in the wake of the preparations for, and eventual accession of, Eastern European countries into the European Union. This closer association of Eastern Europe and the European Union proves particularly troublesome. The recent accession of the 10 Eastern European countries (in May 2004) diminishes the EU's ability to intensify its research and development programs relative to domestic product. These countries historically exhibit substantially lower research spending than the current EU average. There are a number of reasons behind this problem.

In Eastern Europe, even more so than in the Western part of the EU, there is a traditional lack of integration or linkage between academia and industry, due in large part to the extreme autonomy of professors and universities. Even within the technical universities and polytechnic institutes, students are not obliged to participate in industrial internships at enterprises. The professors themselves often enter the education sector directly after their studies. Many professors spend their whole life in academia without much industry experience. Furthermore, industry commits few resources to practical R&D. As *World Trade Magazine* tells us, private spending in Eastern Europe R&D is not very common. There is simply "too much reliance on the state [and even] a perception amongst [Eastern European] business persons that R&D is irrelevant" [12].

While Eastern European countries do conduct fundamental advanced materials research at the university level, Eastern European R&D is quite distant from commercial considerations or possibilities. In fact, innovation is not seen as a convincing competitive strategy in the region (as well as in parts of Southern Europe). Rather, offering low wages and cheap land to the Western countries, Eastern Europe sees foreign direct investment in the traditional industries as key to growth, at least in the mid-term. The fallout from all this is a decaying research and development structure. Since the late 1980s, the number of Eastern European personnel employed in R&D declined significantly. In 1987, every thirteenth person with a higher education was employed in the R&D sector; by 2001, this figure was every twenty-sixth person. For example, in Poland, less than 17% of industrial firms designed or introduced one new product or innovation between 1998 and 2003 [13]. Eastern Europe drags down research and development on the European Union as a whole and actually decreases the European Union's total R&D expenditures relative to its gross domestic product from 1.94% to 1.87%. The policy of subsidiarity then siphons off critical resources from the potentially more creative projects in the West to support innovation in regions where there is yet no convincing structure for it. The result is the overextension of resources across Europe, watering down of potentially robust economies of specialization, and the stagnation of R&D in advanced materials across Europe.

Structural Problems IV: The Coordination Question

Finally, there is a coordination problem. As discussed in the last chapter, more than at any time in the past, advanced-material technology is multidisciplinary in nature. This is a hallmark of modern materials research, which must combine and coordinate physics, chemistry, electronics, biotechnology, as well as other disciplines. The United States excels in undertaking successful, multidisciplinary R&D programs in advanced materials, but European research does not easily cross disciplines. One important difficulty faced by Europe, despite the growing trend in globalization, is the still prevalent regional (and national) biases that limit companies and countries within the European Union from reaching out to outside and complementary firms to form global alliances in joint R&D initiatives. In addition, funding mechanisms within the European Union and nationally are not well designed to manage advanced technology endeavors that are multidisciplinary in nature. Because of the segmented and rigid structure of higher education, peer-review referees do not readily support research proposals spanning different fields. In the United Kingdom, for example, funding comes from different Research Councils, each organized into one area of specialization (e.g., materials). If the subject of the work requires a multidisciplinary approach, the referees of the proposed R&D project may be inappropriate to assess the validity of the proposed investigation, especially if parts of it fall out of the area of expertise of the reviewer. This often results in rejection and denial of funding.

European authorities understand the problem, especially regarding nanotechnology and advanced materials. In 2004, the E.U. Commission stated that in order to "stimulate . . . nanotechnology applications and to capitalize upon the interdisciplinary nature of nanotechnology R&D, it is important [to coordinate] different disciplines . . . in a way . . . to ensure critical mass in applied R&D and to mix different scientific competencies" [14]. Recent programs such as COST (Cooperation in the Field of Scientific and Technical Research) and the PHANTOMS Network attempt to advance multidisciplinary R&D in nanotechnology and advanced materials.*

However, COST and PHANTOMs still struggle with serious problems. They, for example, do not support research itself. It remains difficult for newcomers and the smaller creative firms to break into the COST and PHANTOMs networks and to get clear and concise information about the networks currently running. There is, to date, no tangible evidence that these programs have led (or are leading) to significant practical results.

THE EUROPEAN DILEMMA IN R&D III: THE "CULTURAL" PROBLEM

The structural problems embedded within the workings of the European Union are a relatively recent phenomenon. But we know that as early as the 1850s, Europe already began conceding, if not in any formal way, then certainly de facto, leadership in the industrial revolution to the United States. The structural problems of the present-day European Union are, in fact, a consequence of earlier, deeper sociocultural currents that engendered the uniquely conflicted relationship that Europe has maintained with science, innovation, and technology creation since the midnineteenth century.

Certainly, the cultural, political, and historical peculiarities of each of the European Union's countries, not to mention the international conflicts within Europe that nourished the animosities and bloodshed that the European Union is suppose to

*The COST program stimulates the coordination of European-wide research activities through the funding of meetings, workshops, joint publications, and short-term scientific missions. The COST program helps to form wide-ranging R&D initiatives linking researchers, institutions, and companies. The PHANTOMS Network is a multinational networking initiative. It brings together capacities and talents throughout the E.U. regions and stimulates specifically commercial "nanotechnology applications." The PHANTOMS Network is funded by the E.U. Commission. It is designed to foster interdisciplinary research in the sense that members of the network come from different fields in industry and academia. PHANTOMS includes 200 participants from over 20 different European (and non-E.U.) countries. The network links different research activities in different fields across a variety of regions and countries. It promotes exchange of information and knowledge and develops multidisciplinary network is of European universities and companies.

eradicate, hinders coordination and harmonization of national research and development agendas. Above and beyond these considerations is the historical–cultural relationship that exists between academia, industry, and society in Europe. This relationship within the European context strongly conditions the role of science vis-à-vis technology, the strength and types of linkages between centers of research and the market, and the general attitude of European society toward technical innovation that determines the rate and even direction of diffusion of productivity-enhancing technology within and between the E.U. countries.

Cultural Bias and the Science versus Technology Issue

Europe traditionally excels in advanced science. For centuries, its universities contributed to the most important advances in the physical and biological sciences. The last century was arguably Europe's greatest in the realm of science. This is especially true in those scientific areas directly related to advanced materials: atomic structure, molecular architecture, quantum theory, chemical interactions, and so on. In every case, the great twentieth-century advances in these sciences originated in Europe, particularly Germany, England, and France. The dominance of European science even today comes through clearly in its aggressive research programs in particle physics. Europe is now investing twice as much as the United States in particle physics research. In particular, Geneva is the site of the large ($4 billion) Hadron Collider, seven times more powerful than any American particle accelerator. American research scientists, seeing the writing on the scientific wall, are now leaving United States to take up academic and research positions in Europe.

It is no surprise, then, that the universities in Europe with the greatest reputations traditionally focus on the theoretical sciences. The prestige of European universities depends on the output and quality of science produced by faculty, as indicated by such measures as number of articles published and number of citation of articles by other publications. The European Union as a whole overshadows the United States in terms of scientific performance in nanotechnology and advanced materials. The United Kingdom, for example, remains a hotbed of scientific inquiry within the European Union. With only 1% of the world's population, the United Kingdom carries out 4.5% of the world's science research, produces 8% of all research papers, and wins 10% of the international science prizes. Within the United Kingdom, Cambridge University and its Cavendish Laboratory has traditionally excelled in the pure sciences, and is today Europe's premier institution for advanced-materials scientific research. According to the European Union's "Science and Technology Indicators 2003" report, the European Union as a whole accounts for 34% of the world's nano- and advanced-materials scientific publications, mostly generated within universities. North America's scientific output pales by comparison. It accounts for less than 25% overall. As the United States continues to outsource its R&D, it will likely continue to lose ground in scientific output to Europe.

Historically, in Europe the "manufacturing arts" were the dominion of the lower classes, while higher education was restricted to the aristocracy. Indeed, the use of pure theory for enabling technology was seen as the "prostituting" of science, tantamount to crossing social boundaries, and so considered anathema to European traditions. Certainly, the creation at the end of the nineteenth century of technical colleges, such as Germany's Technische Hochschules, helped to promote industrialization, but their influence never extended beyond the regional, and they were hardly the type of schools elite German and British intellectuals aspired to. The alliance of such universities with local governments to achieve regional economic growth, which did in fact take place sporadically, flies in the face of how European social elites (incorrectly, we now know) viewed economic progress: the central governments were supposed to support "big science" to enable a large-scale industrial revolution, the benefits of which, in the form of economic growth, would trickle down to the masses over time.

The United States presents a different and more supportable model of economic progress, one reflected in the most recent scholarship in the history of technology. Engineering fueled American economic growth and ushered in the "American Century." American engineering has been less occupied with knowledge for knowledge's sake and most consumed with building objects that have economic use in society. In engineering design, practical results control the daily activities of personnel. The engineering designer does whatever it takes to achieve results, often sidestepping science totally and creating new machines and processes using ad hoc empirical methods. Engineers also avoid publication of their results lest they release industry secrets to competitors. The value of engineers' work depends on the number and importance of their patents and the economic value of their creations to working companies, their stockholders, and investors. Simply put, American society historically honors practical application. Some of the most prestigious universities in the United States, including MIT, Caltech, and Virginia Tech, cherish the engineering disciplines. Recently, these universities established close ties to advanced-materials firms, something that has traditionally been frowned upon in European academia.

If European science pushed the realm of pure knowledge further than any other region over the last one hundred years, the commercial value of this science is questionable. A constant and gloomy theme that jumps out of virtually all government reports on Europe's role in the advanced technology industries, especially nanotechnology and the new materials, is that relatively little of Europe's excellent science ever crosses over into industrial practice. European Union officials appear to understand that if Europe is to become industrially competitive with the United States in advanced materials, it must not only devote more of its GDP toward creative research and development. It must also start shifting its R&D away from government support of "big science" and toward practical engineering undertaken by competitive private industry. The United States is looking to reach the point at which a full two-thirds of new R&D investment comes from the private sector. According to the E.U. Research Commission, it is only in this way that Europe can have any chance to boost its competitive potential in the global marketplace. One promising trend is that three countries in the European Union, including Germany, now exceed the United States in the share of total R&D financed by industry. Such was not the case just a few years ago. But does this fact in itself indicate a rising European tide in technology? Not necessarily. Large firms with a scientific bias continue to call the tune and the European Union as a whole still falls significantly behind the United States in indicators of actual technological progress.* It is still uncertain whether the European Union can fully embrace such a model that runs counter to Europe's deeply imbedded science-oriented culture.

Cultural Bias, Public Perception, and Fear of Technology Issues

The perception of, and attitude toward, the taking of risk separates the United States and Europe in how each views new technology. How a country or region perceives risk, in turn, creates an environment that either facilitates or hinders the creation, diffusion, and adaptation of technology. Whereas the United States tends to embrace the risks inherent in innovation and its potential for society, Europe takes a more cautious, even skeptical approach toward the uncertainties that must attend technological development. As Craig Storti tells us in his work on culture conflict between Americans and Europeans, "Americans are notorious for taking risks . . . while [Europeans] find no inherent value—and a great deal of inherent bother—in risk taking. . . . For [Europeans] trial and error are *at best* something to be avoided whenever possible" [15]. Fundamentally, these divergent views derive from how differently the United States and Europe withstand and even embrace what is an inalienable part of risk: the possibility of failure. For their part, Europeans tend to fear and avoid failure whenever they can. Americans, imbued with a "frontier" mentality that considers all things possible, accept and, even more, heartily welcome risk as a necessary step in the learning process that leads to final success. Recent investigations by the European Union support this position. That 1999 report on biotechnology clusters within the United Kingdom previously mentioned found in the United States a "can-do" mentality that has contributed so much to U.S. economic success. The report found evidence that this was "key to many of the achievements in clusters such as Boston and San Diego" [16]. Comments by British entrepreneurs working in the United States point to this factor as the main reason they decided to work in America. British researchers encountered a "refreshingly positive attitude to failure where fear of failure is lower in the United States" [17].

*In Germany, there is some integration between industry and academia via the Steinbeis model for technology transfer. The Steinbeis Foundation operates in the Baden-Wurttemberg region. It is nonprofit organization for technology transfer. The Foundation has over 200 centers, The Steinbeis Technology Transfer Centers, located in polytechnics and universities. These are focused on small, practical, technological problems of small enterprises and companies. These problems are often not on the radar screen in national and even regional R&D support schemes. Managers of the centers are also professors who have industrial experience. Contract research is allowed to supplement income. Each center is its own administrative and profit center, funded from fees obtained from doing work for industry.

European skittishness in the face of "the new" reveals itself in very tangible ways that closely touch advanced materials. Europeans perceive new technology as an uncontrollable force, one posing great risks if unleashed on the world. Caution and further study are the watchwords that greet innovation, especially of the more radical variety. The recent trade dispute between the European Union and the United States involving genetically modified organisms (GMOs) for food crops illustrates the point. Developed by American companies in the 1980s, the U.S. government worked closely with the chemical and agricultural industries to help the diffusion and adoption of GMO technology within the United States. As a result, the American agricultural industry embraced the use of genetically altered seeds, resulting in productivity increase in the growing and harvesting of crops. The European Union differed sharply in its position on and actions regarding GMO seeds, expressing concerns that GMO seeds might cause unforeseen damage to nongenetically modified crops through cross pollination, causing unpredicted consequences. This reluctance on the part of Europe to use this advanced technology conditions its attitude to other novel materials, particularly those associated with nanotechnology. The number of reports and issue papers coming out of Geneva on the potential dangers to the environment and humans of a society increasingly dependent on, and thus exposed to, new and unfamiliar species of nano-materials bodes ill as to Europe's timely acceptance and adoption of the latest technological creations.

This innate distrust of new technology within the European Union, the vestiges of the ancient condescension of "upper-class" science toward corporeal innovation, and the prevalence of a "science-push" model, means that Europe's "scientific–industrial–government" complex maintains a significant distance from the realities of the marketplace. As a consequence, lines of communication between policy makers, scientists, and industry on the one side and the public market on the other rupture and break down. This failure reveals itself in a palpable way in the inability, or unwillingness, of European universities and businesses to publicize their achievements on a broad scale. Whereas, leading American newspapers are not shy to hail U.S. achievements, the European press behaves in a more reserved manner and does not "shout" about major advances. This reticence is linked to Europe's historical reluctance to publicize scientific works. With science controlled by the elite, the public as a whole did not, nor were they expected to, participate in the proceedings of such lofty, specialized work. In 2004, The European Union appeared increasingly concerned with this "disconnect." They have, at the least, expressed in no uncertain terms that investigations by the European Union reveal clear evidence of a barrier to communications between the elite of science, industry, government, and the public marketplace, and that "without a serious communication effort, nanotechnology [R&D] could face an unjust negative public perception. An effective two-way dialogue is indispensable . . . the public trust and acceptance of nanotechnology will be crucial for its long-term development . . . it is evident that the scientific [and technical] community will have to improve its communications skills" [18].

DISCUSSION

Too often in discussions about research and development, the two actors (R&D) are seen as tied at the hip and, even more troublesome, are considered completely fungible entities in the technology creation process; if a company or nation progresses, it does so through an equal competence in, and ability to straddle and coordinate, both realms of human endeavor. But from this chapter we conclude that this is not the case at all. We see, first of all, that research and development in Europe is not the same sort of animal that it is in the United States; Europe embraces the "R" in all of its pure, theoretical glory, whereas the United States focuses on the "D," hewing closely to its ethos of functionality and a "damn the science, markets at all cost" mentality. These different ways of looking at progress are embedded within, and find tangible form as part of, the different structural components that impinge upon the R&D process. We have seen this in particular within the European Union and how it as an organization functions today. But these structures themselves evolve out of a society's inherent and deeply rooted cultural and historical context.

In this age of tolerance of and, even more, appreciation for the beliefs, traditions, and structures of different cultures, it would be satisfying to report that a happy symmetry exists here; the United States and Europe have just found different but equally effective paths to one common goal: economic progress. The goals may indeed be the same, but the routes taken by the United States and Europe to them, though they may be philosophically and morally equivalent, are not at all of the same weight practically. The fact is, the U.S. approach goes much further than that of the European Union in mitigating and even confounding a number of important risks to economic success that we have outlined previously. Certainly, its emphasis on praxis over theory is central. From this flow a number of other characteristics of American R&D. Its "closeness" to the market follows naturally from its innate practicality; a densely structured but efficient feedback system exists between the centers of R&D and the marketplace. This means that American research and development continuously sends out to, and receives signals from, demand sectors so that the final output, whether product, process, or service, conforms closely with the present or future needs and desires of consumers.

The case of research and development within Europe uncovers a starkly different world, one in which the academic scientist stands dead center within the frame, controlling the actions of the characters and defining the perspective of the scene. Deeply rooted in its "big science" mentality, Europe favors that model of industrial competitiveness that embraces central government and the large R&D corporation with a known history and national profile. These firms are seen as national champions whose research activities must be supported and whose scientific research output should then be injected into an unsuspecting, often uncooperative, market. In various ways, we have seen how this predominant trait leads to a progressively greater distance between centers of R&D and the marketplace. From the university system to the European Union as a whole, the structure of R&D in Europe places barriers of various gradations of penetrability between those few who conduct research and the many who in aggregate make up the potential market for the fruits of this creative effort. The "big-science" and "science-push" models that influence the management strategies of European government and industry attempt to create from theory new technologies for a still-to-be-determined market. We have seen from our case studies that this Descartean approach often leads to dead ends, since the complexities of theories do not easily translate into useful results.

It is left to the more practical-minded United States to pick up the loose ends and, always with a skeptical eye as to the usefulness of science, create technology through a bottom-up engineering design approach. And even those technologies created within the United States do not necessarily find friendly markets within the European Union. Fear of competition is but one, and not necessarily the most important, reason for this cold reception. Other, more deeply imbedded fears of the unknown and what "it" may turn loose on society are a far more potent deterrent to "the new." A supreme irony operates here. In its R&D activity, the United States both welcomes the challenge of risk and, at the same time, thinks and acts in ways that minimize and even thwart the risks of creation. Europe, on the other hand, intensely fears risk (and, ultimately, the failure that may come from taking risks) but thinks and acts in ways that create greater risks, even as it vainly searches for those desired avenues that lead to the promised land of economic progress.

We have so far examined the research and development process in the United States and Europe as if it takes place in a vacuum. We have very little sense of the institutional and informational networks that connect R&D to an economy as a whole. There must exist far more extensive and profoundly interactive communication systems linking the main actors in technology creation: universities, government, manufacturers, and users. This is to say that innovation, and the economic growth attending it, not only has a technological dimension, but also a social one. An important reason for the innovation gap exhibited by the European Union is neglect of these social dimensions in the European Union's R&D programs and in the policies surrounding them. Modern technology must be accompanied with complementary new investments in the qualifications of employees, new forms of work organization, new production concepts, and new systems of management. Without considering these social components, real bottlenecks appear to thwart the introduction and efficient utilization of new technologies. Most importantly, creators of technology must establish networks of communications with strategic social groups who help create acceptance and ease entry and distribution of the technology through the economy. The United States is again more advanced in these ideas and techniques than the more tradition-bound European Union.

By fleshing out these greater details, and this broader social context, we can better understand the wider structure of the advanced-materials industry and how its technological output diffuses into and becomes an active part of society and an engine for economic expansion. From this wider perspective, we deepen our understanding of the risks that are an inalienable part of a new product's life cycle and the relative skills with which the United States and Europe manage these potential barriers to technical and economic progress.

REFERENCES

1. Commission of the European Communities (1999), *Towards a European Research Area,* Communications from the Commission (Brussels; European Union Publication), p. 1.
2. Ibid.
3. UK Minister for Science (1999), *Biotechnology Clusters* (London), pp. 30–31.
4. Commission of the European Communities (2004), *Towards a European Strategy for Nanotechnology,* Communication from the Commission (Brussels; European Union Publication), p. 8.
5. Commission of the European Communities (2003), *Third European Report on Science & Technology (S&T) Indicators* (Brussels; European Union Publication), p. 1.
6. Ibid.
7. Huang, Z., Chen, H., Yip, A., Ng., G., Guo, F., Chen, Z. K., Roco, M. C. (2003), "Longitudinal Patent Analysis for Nanoscale Science and Engineering: Country, Institution and Technology Field," *Journal of Nanoparticle Research,* Vol. 5, Nos. 3–4.
8. Ibid. Refer also to Cientifica (2003), *Nanotechnology Opportunity Report* (2nd Edition), Executive Summary; *Nanoinvestor News* (2001–2005), *Facts and Figures;* Salvatore, D. (1999), "Europe's Structural and Competitiveness Problems and the Euro," *The World Economy* (Oxford, UK: Blackwell Publishers Ltd.), Section 3. Also refer to this chapter, Reference 1.
9. Nanoforum (2007), *Nanotechnology in Europe—Ensuring the EU Competes Effectively on the World Stage, Survey and Workshop,* Final Report (Dusseldorf, Germany), pp. 36, 38.
10. UK Minister for Science (1999), *Biotechnology Clusters* (London), pp. 32–33.
11. Nanoforum (2007), *Nanotechnology in Europe,* p. 36.
12. "Eastern European R&D Spending is Small Change" (2003), *World Trade Magazine Online* (www.worldtrademag.com), December 1: p. 1.
13. Ibid.
14. Commission of the European Communities (2004), *Towards a European Strategy for Nanotechnology,* p. 11.
15. Storti, C. (2001), *Old World/New World: Bridging Cultural Differences—Britain, France, Germany and the U.S.* (2001) (Yarmouth, ME; Intercultural Press): pp. 110–235.
16. UK Minister for Science (1999), *Biotechnology Clusters* (London), p. 31.
17. Ibid.
18. Commission of the European Communities (2004), *Towards a European Strategy for Nanotechnology* (Brussels; European Union Publication), p. 21.

Part Four

A Wider Context: The Seamless Web

Chapter 7

Seamless Web I: Companies (Large and Small), Universities, and Incubators

If, as we claim, advanced materials lie at the very center of twenty-first century commercial technology, then it is clear that the United States remains the world's technological leader. Certainly, Europe is today very active in advanced R&D, as significant government money devoted to this testifies. Also, European universities and research institutes explore some of the most interesting and important areas of advanced-materials science. But it appears that it is the United States that most effectively translates scientific and technical discovery into commercial products and processes. Even though Europe increasingly reflects a large, unified market economy, there appears to exist a far closer relationship between American R&D and the marketplace than within the European context.

One perspective in which to understand continued leadership of the United States in productivity through new technology creation is that it has been more successful in managing the inherent risks involved in technological progress than has the European Union. Europe appears to be at a significant disadvantage, even as it grows in power and influence. Structurally and culturally, it takes routes to new technology that extend the time for completion and indeed places roadblocks in the way of successful market placement. The EU's "top-down" approach, in its bias toward the role of central government, the embracing of "big science," and the control of theory over practice reflects a gap or distance between knowledge and application, between the idea and the market.

This and the following chapters deepen our understanding of how the United States fills that gap, that is, bridging knowledge, innovation, and the marketplace. Using advanced materials as their touchstone, these chapters explore in greater detail those pathways by which engineering research flows into the economy. In doing so, we find an intricate and virtually seamless network of old-guard corporations, newer and highly innovative small- and medium-sized start-up enterprises (SMEs), the engineering and materials-science departments of universities, and a growing pool of incubators that effectively link, as intermediaries, industry and academia. The creation of this web, we can argue, is relatively recent. Although the large corporation has been part of past R&D, its role has been receding to secondary status. The rise of the small firm as innovator reflects the dispersed, continuous nature of technology creation in the late twentieth and twenty-first centuries. Indeed, the start-up SME has replaced the centralized, tightly integrated R&D corporation as the prime source of new-materials technology. This greater dispersal of R&D function throughout the economy implies that other institutions—universities and incubators—take on greater importance in the practical world of technology as they become more interconnected with the small- and medium-sized firms to form chains or links of innovation and commercialization that extend continuously throughout the economy. Through such intricately designed networks, the United States, efficiently and in a timely manner, speeds the creation and implanting of new, productivity-enhancing advanced-materials technology into society.

The United States remains quite ahead of other countries in new technology, but not because of government aid and the country's sheer concentration of physical and financial resources, as is often claimed. Nor is Europe's apparent difficulty in pioneering the most critical technologies the result of government inattention, inadequate funding, or poor science. As we have seen, all these resources, in fact, exist in Europe.

What we can say is that in Europe there is a low level of the essential dialogue between the realms of research, development, and the marketplace. There is not in Europe that medium of extensive, continuous, information networks through which technology moves and evolves from an idea on paper to real-world systems. Within the United States, on the other hand, this seamless web and the critical information transfer that goes on within it, and thus the dialogue that occurs between research, development, and the market, appears quite robust and expansive.

A vital function of this creative network is in selecting from the massive output of R&D taking place at all levels throughout the country those most promising projects that seem likely to succeed as creatures of the market, and in rejecting those merely interesting ideas (and patents) that look to be commercial dead ends. By doing this, the seamless network prevents industry from wasting precious time and resources, and thus retain its competitive edge globally. In this and the following chapter we investigate the most essential components of this seamless web: the corporation, SMEs, the universities, incubators, government, and venture capital. In doing so, we delineate their individual natures as well as their vital interactive dynamics.

THE CORPORATION: CAPTIVE MARKETS AND THE LARGE INTEGRATED FIRM

The large firms continue to contribute to the high-technology field, and thus to American economic growth, sometimes in significant ways. In these cases, they support research efforts in creating an ongoing and expanding pool of intellectual capital. Within the advanced-material field, internal or captive markets often play a vital role in narrowing down the types of commercialization efforts that integrated firms undertake. General Electric, DuPont, IBM, Raytheon, and other major technology firms tend to explore those advanced-material technologies that they can use in-house or within the manufacturing facilities of subcontractors. In a limited sense, then, these captive markets serve as selectors and filtering mechanisms for technologies pursued by certain large, vertically integrated companies. There are three types of "captive market" mechanisms at work in these cases.

The "Immediate Captive Market Permanent R&D (ICMPR&D)" Firm

In this case, the firm maintains a large, diverse, and permanent R&D department that continually explores various long-range approaches related to the company's product line and accepted market strategies. At any one time, the firm closely heeds immediate market signals or related internal bottlenecks in order to decide which of the myriad of R&D projects with which it is involved will be given priority in terms of time, resources, and marketing attention. These signals come from internal products and technologies that significantly benefit from the new advanced material. We refer to this type of firm as the "immediate captive market R&D firm" (IMR&DF).

If this type of model is one of the more important ones within the large-firm category, it is also one of the oldest. The history of American technology is replete with such cases. In the 1920s, DuPont created one new material—methanol—in part as a way to break through the roadblock to making another material—synthetic ammonia—efficient enough to be commercially attractive. Since then, methanol production has been closely linked with ammonia (and other related compounds) in petrochemical production. Even more famous is the Niagara Falls industrial complex of the early twentieth century. This early cluster established a teeming network of materials and technologies linked by a complex web of diverse but codependent firms. The adaptation of electrochemical processes for chlorine to make aluminum is but one of numerous examples of technological cross-referencing and interdependence.

But this model of advanced-material creation is not just of historical interest, for we see it at work today as well. The case at General Electric (GE) illustrates this. GE, one of the great pioneers in institutional R&D, established its nanotechnology program in the 1990s in order to investigate and develop diverse technical routes to useful advanced materials. Those materials that serve internal GE products receive the most attention and encouragement. The company particularly looks for new materials that can be applied to improving the performance of such in-house products as aircraft engines, medical equipment, electric appliances, electronic equipment, and television technology and services. In a recent example, GE began to focus on developing and commercializing one particular nanotechnology material out of a number it was investigating. It pursued a three-year, $5.8 million grant from the National Institute of Science and Technology (NIST) to develop a template-synthesis platform for growing large arrays of aligned nanorods. The project focused on making the materials in a precise and controlled manner on a large scale for the company's highly specialized and market-driven products: medical imaging, fluorescent lamps, flat-panel displays, energy, and biomedical (detection of diseases, cancer).

The German automaker Daimler Chrysler entered into nanomaterials for a similar reason as GE: to improve its immediate and economically important operations and products. In its recently conceived "Nanocar" program, the company explored the field of advanced materials for possible uses. It took particular interest in ultrathin coatings as a way to capture competitive advantage in the market. One of the company's newest cars features "Conturan" glass with an ultrathin layer of antireflecting coating. Daimler is also looking at nanocoating technology to satisfy consumer demands for flexible color modification capability through the incorporation of "changeable color" techniques using nanoparticle coatings activated by an electric field.

Another example of the ICMPR&D firm is Xerox. The company has been heavily involved in advanced-material R&D since the 1980s. One of its areas of interest are those nanomaterial possibilities that might be of use to improving its most important internal products or production processes. For instance, in the late 1990s, Xerox faced the problem of having to economically make smaller, more precisely structured copy-toner particles. This presented a technical problem that needed to be resolved if Xerox was to ward off rapidly growing competition in the toner field. Accordingly, the Xerox Research Centre (in Mississauga, Ontario, Canada) targeted a line of research to develop a technology called the emulsion aggregation (EA) process to create toner-quality nanoparticles. The smaller particle size produced by this process gives much improved image resolution and the economic advantage of having to use about 40% less toner on a page. The process also allowed Xerox to better control the structure of the particles and the ability to imbed within the toner particles themselves various dopants for improved performance and multifunctional capability.*

The "Immediate Captive Market Ad Hoc R&D (ICMAHR&D)" Firm

This model is similar to, but distinct from, the ICMPR&D firm. In this case, the firm does not maintain a large, diverse, or permanent R&D department. Historically, these firms inhabit industries that do not rely heavily on large-scale research and development for market growth. There is no extensive and deep R&D pool that it has developed over many years from which to draw upon. But they get drawn into undertaking short-term R&D by the immediate signals of the marketplace. In the most common case, they lose their competitive edge to more technologically advanced competitors. They then rush to find a way to gain back market share. Min-

*In a similar vein, large chemical and information technology (IT) companies also appear to embrace the ICMR&D model. For example, the chemical firm BASF pursues advanced nano-structured water-repellent coatings to improve the surface properties for its current coating process.

ing the advanced-materials field to bolster their existing product line seems to them to be their most promising route. They then undertake a limited, temporary research effort to deal with the specific problem at hand. Rather than investigate many possible alternatives, they grab onto the quickest way to solve their immediate technical or market-related problems.

The model for the ICMAHR&D firm is Burlington Industries. Not known for taking on extensive and leading-edge R&D—the large chemical firms in the 1950s, 60s, and 70s accepted the responsibility for developing new fibers—Burlington Industries only recently began specific and short-term research and development activities to solve a difficult market situation. Started in the1920s in North Carolina, Burlington made good use of locational advantages and organizational and marketing strategies to become, by the 1960s, the world's largest textile company. At its height, Burlington controlled 140 U.S. textile plants and closed in on $4 billion in sales. But by the late 1990s, the company had reached a critical point. Profits, which stagnated through the 1990s, now fell rapidly. These difficulties resulted from low R&D activity over the years and the flood of low-cost foreign textiles from more technically advanced and strategically savvy international companies in the wake of the globalization movement. By the late 1990s, Burlington attempted to adapt specific nanotechnologies into its existing fibers in order to obtain competitive advantage through differentiation strategy. The success of this approach still remains to be seen.

Large firms in other industries, historically perceived as "low-tech," increasingly must play catch up as their competitive positions weaken under the pressures of global competition. These firms then conform to the ICMHR&D model; they include companies from the metal making and fabricating industries, the assembly industries, mining, petroleum production, and related sectors.

The "Projected Captive Market Permanent R&D (PCMPR&D)" Firm

Large firms, especially those with diverse income streams, might not just target R&D projects in response to *immediate* market signals. They might also perceive, rightly or wrongly, that a potential market or technical problem lies down the road and start to deploy attention and resources to solve these problems before the fact. Firms juggling a variety of businesses are more likely to do this since they have "cushion money," allowing them to spend time on "on-the-horizon" issues. These firms almost assuredly are of the permanent R&D type because only these R&D departments have the breadth and deep understanding of the field to recognize possible future problems. One of the classic examples of "projected" development activity was in the aircraft industry, with the evolution of the jet engine in the 1940s. Here, development began with economic and technical difficulties predicted down the road for the traditional propeller technology. There now appears a mounting trend toward this type of "predictive" impetus to change in such advanced areas as modern materials.

Within the electronics and semiconductor industries, for example, experts question the future viability of Moore's Law, which, as previously noted, states that the number of transistors in a given area of an integrated circuit doubles every 12 to18 months. Computer processors have become so dense with circuits that traditional methods of designing and making chips are expected to become economically untenable by 2020. It is for this reason that IBM is exploring the potential role of nanotubes in its future integrated circuits as a way to maintain, and even surpass, Moore's Law. IBM finds that, with nanotube technology, molecules arrange themselves into patterns (like snowflakes). This means that individual chip circuits no longer have to be drawn by lithographic means, thus saving on the costs of labor, equipment, and time. Also, with traditional wire circuitry, designers must deal with resistance, scattering of electrons in many directions, and lost energy. In contrast, nanotubes are so long and thin, electrons cannot be directed sideways and so face minimal resistance as they travel, thus considerably reducing energy loss. With the possibility of nanotubes replacing copper conductors and silicon semiconductors in future circuits, IBM looks to override predicted technical and economic barriers.

In a very real sense, the U.S. military itself, through its network of contractors, conforms to this model. Although the military sees its current technology as equal to present day military requirements, such as in the war against terrorism domestically and in Afghanistan and Iraq, it understands that it must do a better job of foreseeing the types of defensive and aggressive weaponry and material that will be required in the future. A case in point is the creation of the Institute for Soldier Nanotechnologies (ISN). Housed at MIT, the ISN and MIT enlist a group of industrial partners, including DuPont, Raytheon, Dow Corning, Carbon Nanotechnologies, and others, to research and develop advanced-materials technologies for use in battle. MIT's Department of Chemical Engineering plays a key role in coordinating information and research between its broadly interdepartmental faculty, the commercial subcontractors, and the government. Advanced-materials research is the central work of the group. The consortium investigates nanotubes, micro magnets, advanced polymers, composites, and other materials to design light uniforms; super-strong and germ-resistant fabrics; enhanced sensors to detect poisons, microbes, and radiation; and advanced bandages and splints for battlefield injuries.

These three different models of focus and selection tend to be used by different types of large firms and organizations in certain industries. The matrix in Table 7.1 relates each focus and selection model to the appropriate type of industry.

THE RISE OF THE SMALL START-UP FIRM IN ADVANCED MATERIALS

In response to the changing nature of global competition, American industrial R&D finds itself shifting from centralized corporate departments concentrated within the United States to decentralized, geographically dispersed activity. Despite the advanced-material activity of certain corporations, the larger, integrated firms are less able to rapidly move and take advantage of new technological opportunities and so

Table 7.1. The large, integrated firm—types of industries associated with the different selection and focusing mechanisms

	High-technology industry	Low-technology industry	Military-related industry
ICMPR&D model	X		
ICMAHR&D model		X	
PCMPR&D model	X		X

tend to put the brakes on radically new development. Also, the increasing costs for R&D, which is a very uncertain activity, and the difficulty of creating products that will compete with established materials, dampens the enthusiasm of traditional chemical and material corporations for exploring new avenues.

In sharp contrast, SMEs have actually taken the lead in innovation, especially in new-materials work. Recent research testifies to the expanding role played by SMEs in creating and diffusing those new technologies that drive economic progress. Table 7.2 shows the growing percentage of SMEs that create and guide new-material technology into the marketplace. In the 1970s, less than 30% of SMEs spearheaded new-materials innovation. By the 1990s, this number grew to over 50%. It is forecast that by 2030 SMEs will be responsible for nearly three-quarters of such innovations. On the other hand, the table shows a sharply diminished presence of the large corporation, from being at the helm of over 65% of new materials innovations to less than 20% by 2020.

Decline of the Large Corporation as Technology Leader

It can be argued that the self-containedness of large, integrated firms ultimately led to their undoing since they also remained isolated from the outside as their organizations matured and rigidified from within. By the 1960s, these "old-guard" firms found themselves in turmoil as newcomers more and more competed with the established corporations in formerly protected market segments. A major reason for this shift was the growing influence of such chemical process engineering compa-

Table 7.2. Source of major advanced-materials technology: 1970–2030 (% of major innovations)

	1970	1980	1990	2000	2010	2020	2030
Large corporations	65	50	35	30	25	20	15
SMEs	15	35	50	60	65	70	75
Other	20	15	15	10	10	10	10

Source: [1].

nies as Kellogg, Stone & Webster, and Scientific Design. As Peter Spitz reminds us, by the 1960s, these firms designed and constructed standardized turnkey petrochemical plants. This development, in turn, allowed smaller, upstart firms to enter markets that were previously the preserve of large corporations, and without undertaking significant new R&D. Whereas, previously, one or two companies dominated a particular material field for example, nylon, polyethylene, and so forth, now a half dozen smaller firms competed. Also, with the engineering firms exporting their technology overseas, Europe and Asia challenged the United States in international markets. The heightened competition in the industry led to periods of overproduction and reduced profit margins. This meant less money for major R&D ventures. The oil shocks in the 1970s further decimated the petrochemical chemical industry.

By the 1960s, the large established chemical and petrochemical companies faced internal resistance to undertaking new and challenging technology. The salaried, professional managers preferred company policies that favored the long-term stability and growth of their enterprises to those that maximized current profits. This more cautious approach preserved the viability of the company without the uncertainty of risk. The trend toward professional specialization also had a hand in the slowdown in innovation. As managers specialized, they each took on their own priorities, goals, and even languages. Communications between the various functional departments within an organization weakened, thus blocking the smooth flow of information between departments. The critical coordination between supply, transportation, production, marketing, and distribution suffered accordingly, and the costs of putting in place a new product on a mass production basis increased prohibitively.

This evolving professional manager, who was more and more likely to be a graduate of a leading business school, developed strategies based on sophisticated quantitative models biased toward short-term, financially safe strategies. These leaders in informal and sometimes formal alliances with the banking community, saw greater advantages, especially in mature industries, of quick profits from divesting units than from long-term R&D. By the 1970s, leveraged buyouts and debt reduction became the major dynamics in industry. Aftalion [20], for example, points out that DuPont, Monsanto, ICI, PPG, and others divested by selling petrochemical units to corporate raiders such as Sterling Chemicals, Vista Chemical, and Cain Chemical. These smaller, individual "entrepreneurs" could not enter the chemical business except through purely financial, rather than technical, means. The large chemical companies, once the giants of their industry, now became smaller and refocused on specialty chemicals and products for which they were not necessarily adapted or suitable. Such specialties, such as the energy sectors, did not usually thrive.

More than ever, businessmen active in buying and selling companies did not even attempt to operate the firms they purchased but rather bought up large corporations with the aim of selling off their parts piece by piece, thinking that the individual entities of these chemical conglomerates held greater value than did all the operations together [2].

All this led to either the diminution or disappearance of once innovative companies. Even firms that tried to fight these trends lost out because they had to borrow great sums of money to ward off the attempted hostile takeovers which, in turn, forced them to sell off their best and historically most productive and innovative units to pay off the debt. Aftalion reminds us that this happened to Union Carbide, which has not yet recovered. Other victims in one way or the other of corporate raiders were Uniroyal, Phillips Petroleum, Celanese, Stauffer Chemical, and Allied Chemical.

As a result of these economic, organizational, and managerial shifts, the once creative materials firms, those large integrated petrochemical companies, began shying away from the more risky ventures and increasingly adopted incremental product differentiation and improvement as viable and more certain market strategies. In effect, the industry had reached a point of extreme maturity to the extent that few if any major discoveries could be expected to pay off in the near term. The industry did explore such areas as biotechnology and composites but only in an incremental sense, since no major benefits could be expected for decades. Simply put, the best days of R&D within the large materials companies were over. Rationalizing and restructuring in the chemical industry led to painful decisions that had to be made by a new generation of chemical executives. This meant shutting down plants, cutting personnel, and redirecting efforts to areas that were less susceptible to economic fluctuations. Hoechst, Union Carbide, DuPont, Monsanto, ICI, and USS Chemicals divested what were once their most innovative units. Although chemical companies did integrate downstream to petrochemicals, they did so to maintain old technology, not to delve into new areas. Mostly, they wanted to enter into niche markets or felt they could lower input costs due their control over petroleum feedstocks.

This sort of "reduced-risk" strategy permeated the petrochemical "macromolecular polymer" industry into the twenty-first century, and will likely do so into the foreseeable future. Discussions with a number of executives in the chemical industry hammer home this point. The general agreement is that commercializing new polymers today is a very risky and far too costly proposition, and there is no guarantee that there will even be a market for such products. It seems to make more sense to modify those basic polymer products that are already out there today [3]. Dow Chemical, GE Plastics, and other once innovative firms in fact employ the reasonable strategy of exploiting known technology to extend markets. Modifications to the nylons, polyethylenes, polycarbonates, styrenes, polybutylenes, and other well-known polymer materials include improving their chemical, physical, and aesthetic qualities to better penetrate the automotive, electronic, telecommunications, and biotechnology markets. Whereas, in the decade after World War II, R&D departments of the large chemical and petrochemical devoted no more than 30% of their resources to improving or modifying existing products, by the 1990s, this figure jumped to over 65%.

The traditional chemical company—the source of new materials in the past—today exudes complacency and perhaps even disillusionment with investment in innovation. There was definite slowing of investment in innovation by chemical companies in the 1990s, even as the decade brought unprecedented growth to the economy as a whole. Between 1989 and 1998, R&D spending by the major chemical companies grew slowly at 2% annually. In terms of intangible capital, the traditional chemical industry, once one of the most technologically advanced industries, now ranks roughly in the middle of all major industries, lagging behind such sectors as electronics, software, pharmaceuticals, and even oil and gas. Looking at the pharmaceutical industry, for example, R&D spending of the major pharmaceutical companies increased at an average annual rate of 22%. Compared to such technologically robust industries, the chemical industry has become a de facto commodities industry in the manner of steel. These sorts of industries produce low unit value products and shy away from long-term payoffs.

Up to the 1970s, chemical companies within the United States and internationally expanded through their own innovative thrust, often times creating totally new departments centered on new technology. Now, the larger firms grow by merging, acquiring, or joint partnering, often to achieve complementary technologies and product lines. In recent years, for example, BASF extended its reach by acquiring Honeywell businesses, thus complimenting automotive plastics operations and allowing greater penetration of the company's existing plastics—the polybutylene terephthalates (PBT), polyacetals, polysulfones, and polyethersulfones—to U.S. automakers.

DuPont offers another example. It acquired Eastman's high-performance plastics business, which includes a family of polyethylene terphthalates (PETs). The purchase makes DuPont second ranked in the liquid crystal polymer (LCP) area (behind Ticona). These materials were already known in the market, especially in electronic applications. By this acquisition, DuPont, without original research on its own, established its foothold in the so-called "semicrystalline" polymer area. Dow Chemical also pursues a modest approach to new polymers. The company suspended its efforts in developing new types of styrene–ethylene interpolymers and polycyclohexylethylene materials mainly because of rising costs. And Dow is dragging its creative feet in its program involving advanced types of polystyrenes. So, instead of reaching broadly for a number of different potential markets using new technology, Dow modifies impacted technology and finds new market areas in automotive, appliance, and electronics applications. Similarly, British Petroleum (BP) divested its innovative advanced plastics business and staked its future on the expanded marketing of high-density polyethylene, a material on the market for decades [21].

Emergence of American SMEs as Centers of Innovation

We must then look elsewhere for the source of the newest and most important new-materials innovations. As already noted, increasing attention must be paid to the small- and medium-sized firms that are key innovators and the absolutely critical contributors to regional and cluster development. More so than any other area of technology, this is true in the advanced-materials arena. SMEs may conduct R&D as stand-alone companies or they often establish partnerships with larger corporations both within the United States and internationally. The start-up, being the creative unit, generates and transfers the intellectual property essential for creation, while the corporate partners provide financing, market access, and complementary technology [4].

Decentralization of research and development, from a few large companies to many smaller start-up firms represents a development critical to the surge of technological progress witnessed in the 1980s and 1990s. These start-ups emerged to fill the creative vacuum left by the established corporations that no longer pursued pioneering R&D. The founders of these new enterprises saw commercial promise in the research they conducted in academia and created their companies to exploit these possibilities. At first supported by government contracts and venture funding, they eventually progressed to the point where they leveraged their acquired know-how to form partnerships with the well-known corporations and move their technology forward into the marketplace. These start-ups, ever creative, persistent, and flexible, furnish the intellectual energy, the essential ideas and techniques, while the established corporations, at one time the generators and repositories of the most advanced products and processes, now play only a supporting role in the progress of twenty-first century technology and the economic growth it creates. The encroaching of the small start-up firm into the realm of new technology is a phenomenon that is centered in the United States (and to a lesser extent, Asia). Approximately 65% of all start-up firms that came onto the scene between 1985 and 2003 were based in the United States, and by 2015, at least 75% of new SME firms will originate in the United States [5].

The American SMEs not only proliferated within the United States itself, but globally as well. Many small start-up firms leveraged the intellectual capital they established through their research and development activities to negotiate agreements with larger multinational corporations. From 1985 to 1993, U.S. firms increased their investment in R&D abroad three times faster than they invested domestically. Cooperative R&D, in the form of new joint research ventures between firms, is a growing trend as globalization has blossomed as an economic force. This strategy provides additional funding, growth-phase management capability, and access to critical global markets. A typical example is a joint R&D agreement formed between the United States firm Konarka, located in Lowell, Massachusetts, and the French utility industry. The latter is collaborating with Konarka to develop and launch new conductive polymers and polymer-based photovoltaic products into the world market. The utility companies that are working with Konarka in France provide expertise from their European operations to accelerate the development and rollout of Konarka's polymer voltaic products internationally.

In another important case, Quantum Dot Corp. (QDC), founded in late 1998 in Hayward, California to manufacture nanocrystals for biomedical use, entered into a partnership with the Japanese electronics giant Matsushita (Panasonic). The two companies agreed to collaborate on developing tools for DNA detection and other diagnostic applications. The partnership's first product, introduced in September 2003, is a high-throughput optical scanner targeted to the billion-dollar gene-expression-analysis market. In this arrangement, Matsushita developed the scanner and integrated it with Quantum Dot's nanocrystal technology.

An interesting variation of these strategic models involves the sequential spin-off of more specialized firms. For example, Aveka Inc. was spun off from 3M Corporation to develop new ways to process advanced-material particles. Aveka, in turn, split off its own R&D unit, Cima Nanotech, to specialize on researching and developing new processes for manufacturing a particular group of metal nanoparticles. Cima Nanotech then merged with the Israeli company Nanopowder Industries to produce nanoparticles for use in electronics (conductive ink for computer printers and printed circuits) and next-generation rocket fuel (combustible aluminum-based nanparticles).

The field of nanocarbon materials is particularly well known for collaborative efforts due to its multidisciplinary nature, its scientific and technical difficulty, and the expense of advancing this still new field. These agreements follow the business model of linking the small start-up firm specializing in this area and a larger established manufacturer looking to incorporate the new technology into its products. In one case, Performance Plastics Products (3Ps) formed in 2003 a joint development agreement with Carbon Nanotechnologies, Inc. (CNI) to develop new polymer products based on CNI's breakthrough single-wall carbon nanotube technology. These polymers are then joined with 3Ps custom manufactured components for application in industries with special needs in dealing with high temperatures, corrosion, pressure, wear, and lubricity* [6].

UNIVERSITIES, SMEs, AND INCUBATORS

The universities, with their close association with the regional incubator system, today play a more important role in the sociotechnological fabric of society than at any other time in the past. With the rise of the small- and medium-sized firms as centers of innovation, universities have gained prominence as the suppliers of early-stage knowledge and as the life blood of new technology. This is a paradigmatic shift in technological change and industrial competitiveness. Traditionally, such economic factors as natural resources, cheap labor, and demand structure held center stage in explaining why one region or country excelled economically over another. Economists typically explain the rise of industrial complexes around the Connecticut Valley (mechanical production in the nineteenth century), the Niagara area (electrochemical production in the period before World War I), and the Gulf Coast (petrochemical production in the post-World War II period) in the context of superior factors of production.

Historically, the most important roles played by universities in the economy were in educating new generations of engineers and businessmen who entered into

*A start-up company, as it expands, may form relationships with more than one firm, if applications spread in different directions. Thus, CNI in 2003 also entered into agreement with the company NanoInk, a venture-backed firm organized to exploit the commercial possibilities presented by Dip Pen Nanolithograpry (DPN).

corporate America. Indirectly, then, academia helped shape the modern economy. On the other hand, university research has not had in the past too much direct relevance to the creation of new companies and innovations. To be sure, there have been links between the two communities. MIT, for example, pioneered these connections in the early part of the twentieth century through its famous Practice School of Chemical Engineering. This approach proved to be an exceptional way to provide practical experience for students and to train the future generation of engineers and managers. However, students generally learned existing manufacturing technology and did little to advance original research. Further, the gap between the university and industry widened after WW II as the academic engineering disciplines increasingly mimicked the pure sciences in the pursuit of the theoretical and application of advanced mathematics.

The universities did not directly influence industrial change until recently. Never before in the long history of the still-evolving industrial revolution has university-based intellectual knowledge been so intimately linked to economic progress, industrial growth, and their consequences, such as the rise of community development. Within the United States, particularly important shifts have taken place toward a close, interactive, and productive academia–industry alliance.

Universities and the High-Technology SME

More so than in the European Union, close ties between academia and industrial R&D emerged in the United States beginning in the 1980s. Important start-up companies in the field came into being as a result of applied research conducted in universities, particularly in their engineering and materials-sciences departments. The universities are the initial points of contact between early research and commercial technology in U.S. advanced materials. Increasingly, the universities act as creative sources for important advanced-material products. By 2005, over half of the important U.S. advanced-material companies emerged as spin-offs from, or simply licensees of, universities. Many of the most important companies feed off academia as they proceed to development and commercialization. Table 7.3 compares trends in academic-related advanced-material technologies within the United States and the European Union. United States universities began to pull rapidly away from European universities as innovators by the 1960s. The gap continued to widen through 2000 and is expected to continue to grow past 2010. The science itself remains alive and most active within the European academy, but the ability to convert that knowledge into practice clearly remains an elusive goal for Europe's university elite.

A typical case that shows the creative synergies between U.S. universities, corporate entities, and new technology involves the Atlanta-based company Micro-Coating Technologies (MCT). The founder of the company, an academic, obtained his doctorate from the Georgia Institute of Technology in 1994. He based his company on his doctoral research involving an improved method of making thin films. In its first year of operation, MCT had five employees, four of whom had doctor-

Table 7.3. University-related advanced-material technology (% of innovations generated by universities) United States and European Union

Year	United States	European Union
1930	2%	1%
1940	5	5
1950	7	5
1960	8	4
1970	10	4
1980	15	6
1990	30	8
2000	40	8
2010	65	10

Source: [7].

ates, and $400,000 worth of government contracts. At this time, the company generated only $20,000 in private revenue. Over the next few years, MCT expanded its research and development activities until it obtained a fully working commercial process. Growth followed quickly. By 2000, MCT employed over 100 people, including a significant number of production line personnel, and was worth in excess of $200 million.

An important indicator of academia's growing relevance to industry is the fact that, since the early 1980s, American universities took it upon themselves to create windows to the outside markets through university-administered technology management offices, and thereby facilitated technology transfer from academia to the commercial world. These offices serve as management guides to university researchers in taking in-house work and commercializing it. These might be offices of the university or not-for-profit corporations that license university inventions for later commercial use. University technology licensing offices often work closely with venture capital groups to help university researchers to found companies. To take a typical example, Harris and Harris has worked with MIT's technology licensing office to start numerous companies over the last few years (including 25 companies in 2001 alone). A typical university technology management office, such as the one at the University of Virginia, is a private, not-for-profit corporation that evaluates, markets, licenses, and protects inventions generated by faculty. Often, these offices have an in-house patenting department, which reduces dependence on expensive and at-a-distance law firms. For example, The University of Virginia Patent Foundation is a private, not-for-profit corporation that evaluates each of the many inventions generated by University faculty, protects those more commercially promising inventions, and then helps to market and license those rights to industry [8].

Table 7.4 shows the close relationship that exists between the university and innovative, advanced materials SME. Over half of the U.S. companies listed are spin-offs from, or simply licensees of, universities. The majority of the most promising U.S. advanced materials firms feed off academia in this manner.

Table 7.4. University licensees and spin-offs, 2004

Company	Country located	University licenses/ spin-offs	Non-University links	Technology
Applied Nanoworks	USA	Rennselaer Polytechnic Institute		Nanocrystals
Arrow Research Corp.	USA	Cal Tech		Nanocomposites
Carbon Nanotechnologies	USA	Rice University		Nanotubes
Capsulation Nanoscience AG	EU (Germany)		Max Plank Society	Nanocapsules
Cargill Dow	USA			Biosynthetics
Crystal Plex Corp.	USA	Indiana University		Nanocrystal Beads/ nanocrystals
Eikos, Inc.	USA		Internal	Nanotubes
Elan Corp., PLC tals	EU (Ireland)		Internal	Advanced Crys-
Evident Technologies, Inc.	USA		Ioffre Institute (Russia)	Nanocrystals/ quantum dots
Flanel Technologies, Inc.	EU (France)		Internal	Nanoparticles
Hybrid Plastics, Inc.	USA	University of Southern Mississippi Auburn University		Advanced plastics
InMat Inc.	USA		Internal	Nanocomposite coatings
Isonics corp.	USA		Internal	Silicon nanomaterial
Lucent Technologies	USA	Northwestern University		Organic electron-ic materials
Lumera Corp.	USA	University of Southern California, University of Washington		Advanced polymers; Electrooptical polymers
Luna Nanoworks	USA	Virginia Tech		Fullerenes
Monsanto	USA			Genetically modified organisms
Nano C	USA	MIT		Nanotubes
Nano-Tex	USA	University of California–Berkeley	Internal	Advanced fibers
Nano BioMagnetics, Inc.	USA	Oklahoma University		Nanopowders

(*continued*)

Table 7.4. Continued

Company	Country located	University licenses/ spin-offs	Non-University links	Technology
Nanocor, Inc.	USA		Internal	Nanopowders/ polymers
Nanodynamics, Inc.		University of Southern California (USC)		Advanced cramics, nanotubes
Nanofilm, Ltd.			Internal	Ultrathin Coatings
Nanogate Technologies GmbH	EU (Germany)		Internal	Multifunctional materials; nanopowders
Nanomat	USA		Internal	Nanopowders
NanoProducts Corp.	USA			Catalysts/ nanopowders
Nanoscale Materials	USA	Emory University		Metal oxides/ nanopowders
Nanostellar alysts	USA		Internal	Composites/cat-
Nanotechnologies, Inc.	USA		Internal	Nanopowders
Optiva, Inc.	USA		Internal	Thin-crystal film
Oxonica	EU (United Kingdom)	Oxford University		Nanopowders
Polycore Corp.	USA		Internal	Nanocompos- ites; nanoclay
Quinetiq Nanomaterials, Ltd.	EU (United Kingdom)		Internal	Quantim dots; nanopowders
Quantum Dot Corp.	USA	MIT, University of California —Berkeley		Quantum dots
Quantum Sphere	USA		Internal	
Reactive Nanotechnologies	USA	Johns Hopkins University		Advanced bind ing materials; nanofoil
Showa Denko K.K.	Japan			Carbon nanotubes
Southwest Nanotechnologies, Inc.	USA	University of Oklahoma		Carbon nanotubes
SuNyx	EU (Germany)		Internal	Nanopowders
Starpharma	Australia	Bimolecular Research Institute		Dendrimers

Source: [9].

Incubators Enter the "Seamless" Network

The distance between university research and work-a-day technology is often significant. An important institution that has evolved to bridge this gap and help academia-based research become useful technology is the incubator. As varied as they are, incubators add significantly to the continuity and coherence of the seamless web of technology creation within the most advanced economies. They nurture new technology by bringing together specialized resources dedicated to supporting and assisting companies during the innovation process. They typically consist of flexible accommodations, administrative services, consulting advice and linkages to business networks, especially financial ones.

Incubators certainly play a significant role in the research, development, and commercialization of advanced materials. Table 7.5 shows the growing part incubators have had, and will continue to have, in advanced-materials technology within the United States.

There are hundreds of incubators within the United States alone that touch either directly or indirectly on advanced-material development. Table 7.6 identifies and characterizes just a sampling of the major incubators operating in 2005 in the United States on advanced-materials projects in conjunction with engineering schools, private industry, venture groups, and federal, state, and local governments. The data further illustrates the close interactive dialogue taking place in technology creation between some of the major players in academia, business, and government.

The concept of the incubator is relatively new. In the United States, one of the first high-technologyincubators was established in 1959 in Batavia, NY. Beginning in the 1970s, the idea of the incubator gained ground, not only in the United States but in Europe and Asia as well. In the 1980s and into the 1990s, a great many incubators emerged as government projects to attract young companies into a state or region. From the mid-1990s, a growing number of for-profit incubators came into being through the auspices of the private sector: independent entrepreneurs, financiers, large corporations, venture capitalists, and academia.

There are three types of advanced-material Incubators. "Launcher" incubators support start-ups during their early years, with aim of reselling their stakeholdings. "Accelerator" incubators obtain stakes in companies already operating and assist

Table 7.5. Percentage of U.S. advanced-materials innovation impacted by incubators, 1970–2020

Year	Percentage
1970	< 1
1980	5
1990	20
2000	30
2010	48
2020	63

Source: [10].

Table 7.6. Selected major advanced-material incubators in the United States, 2005

Incubator	State	City	Date began	University affiliation	Government affiliation
Ceramics Corridor Innovation Center	NY	Alfred	1992	Alfred University	New York State Office of Science, Technology & Academic Research (NYSTAR)
Long Island High Technology Incubator	NY	Stony Brook	1984	State University of New York at Stony Brook	N/A
Louisiana Business and Technology Center	LA	Baton Rouge	1988	LA State University (LSU)	LA Economic Development Office
Purdue Research Partners	IN		1993	Purdue University	N/A
Rensselaer Incubator	NY	Troy	1984	Rensselaer Polytechnic Institute	New York State Urban Development Initiative
University of Central Florida Technology Incubator	FL			University of Central Florida (UCF)	N/A
SUNY Albany Incubator	NY	Albany	1999	SUNY Albany	New York State Urban Development Initiative
Northwestern University/ Evanston Research Park	IL	Evanston	1995	Northwestern University	
NASA Ames Research Center	CA	San Jose	1992		NASA, Office of Technology Partnerships
Research and Technology Park	WA	Pullman	1990	Washington State University	
Stevens Technology Ventures Incubator	NJ	Hoboken	1990	Stevens Institute of Technology	

Source: [11].

them in preparing for initial funding rounds with venture capitalists. "Holding" incubators take shares in companies that are past their initial funding rounds to develop them and take them to the next level of growth. Clearly, the first type of incubator entails the greatest risks since it is the most likely to take on start-up firms. We can also group incubators with respect to type of sponsor: university, corporation, venture group, or public body (federal, state, or local).

Incubators can be parts or extensions of universities. In these cases, the university uses the incubator to help transfer science and technology from the academic laboratory to the commercial marketplace. Examples of these sorts of incubators include the French Crealys (in Lyon) and Neuvitech (in Ile de France), and ATDC in Atlanta, Georgia (USA).

Corporations also establish advanced material incubators, often as extensions of their R&D departments, in order to gain access to new and emerging technologies, access to new markets and new applications, and as pure investment opportunities. Venture capitalists themselves establish affiliated incubators for a number of reasons: to obtain higher rates of return on a wide variety of technological developments, and to maintain physical proximity to young developing companies, thus allowing venture capitalists to watch over them and, as a result, be better able to detect and withdraw support from "dud" projects, and, alternatively, spot and nurture the future stars. Ultimately, establishing their own incubator group gives venture capitalists a competitive advantage in their field. As of 2000, about one-quarter to one-third of all advanced-material incubators were created by venture capital firms in United States. A leading American venture capital firm that does business with such incubators is Draper, Fisher, Jurvetson. This firm created the Gotham Venture incubator in Silicon Valley and Cambridge Innovation in Massachusetts. (Compared to the United States, Europe registers a high degree of uncertainty avoidance so that European venture capitalists have created relatively few incubators themselves.) Federal, state, or local governments, often coordinating efforts with regional universities, companies, and venture capitalists, create and manage (either directly or indirectly) advanced-material incubators in efforts to engineer economic development and job creation within an area.

Between 1990 and 2001, the total number of incubators in the world rose from about 700 to close to 3000 (see Table 7.7). Although incubators have spread globally since the 1970s, the system of incubators created within the United States is the most extensive, innovative, and specialized. The United States pioneered the creative structuring and application of incubators. As a European report on global incubators concludes, "The United States is surely the country which has generated the most original initiatives creating specialized incubators, dedicated to specific audiences or specific occupations" [13]. Within the United States, the major trends in incubator development are the growing geographical spread of mixed-use incubators, more active involvement of communities in the private and nonprofit management of incubators supported by local organizations, and diversification in the type of incubator toward new specializations.

Over the last two decades, U.S. academic-based incubators have exploded on the scene. In the late 1980s, less than 10% of incubators in the United States collaborated with universities on commercially promising activities. In 1998, this figure had grown to 25% and by 2003 to 37%. The university incubator networks within the United States have become more tightly bonded with the actual marketplace through a number of distinctly American mechanisms. It is little surprise, then, that the most highly regarded academic incubators in the world are in the United States and associated with such universities as University of California , Columbia University (New York City), Michigan State University (Michigan), Rensselaer Poly-

Table 7.7. Distribution in number of incubators, by region, 2001

Region	Number of incubators
North America	1000
Western Europe	825
Far East	600
South America	200
Eastern Europe	180
Africa and Middle East	180
Total	2,985

Source: [12].

technic Institute (New York State), University of Austin (Texas), Georgia Tech (Georgia), and MIT (Massachusetts). Table 7.8 shows the commercial performance (in 1999) of some of the major academic incubators within the United States.

In the United States in particular, we see the nearly seamless interlocking dependencies between incubators, companies, markets, and governments. These interlinkages create vital synergies that accelerate research and development activity into commercial technologies. United States incubators tend to be located in metropolitan areas, often in the heart of industrial cluster groups, with high concentrations of technology-based companies and associated business support firms (e.g., accountants, intellectual property lawyers, and human resource consultants) familiar with the problems of launching and sustaining a high-technology enterprise. Incubators within the United States typically emerge from state and local, as opposed to Federal, government programs that hope that incubator-generated technologies will spur local economic development and job creation.

Table 7.8. U.S. academic-based incubator performance, 1999

Institution	License revenue (millions of dollars)	License revenue (as a percentage of the research budget)	Number of start-ups created (in 1999)
University of California	73	4.3%	13
Columbia University	61	23.6%	5
Florida State University	46	41.6 %	2
Stanford	43	10.8 %	19
Yale	33	11.1 %	3
Carnegie Mellon	30	17.7 %	—
Michigan State University	24	12.6 %	1
University of Washington Research Foundation	21	4.9 %	2
University of Florida	19	7.9 %	2
MIT	18	2.4 %	17

Source: [14].

These and similar programs promote innovation by bringing together in a closely knit network universities, state and local governments, and private-sector firms. In such an integrative manner, the U.S. incubator serves to facilitate the creative dialogue between ideas, technology, and markets. Within these "bottom-to-top" societies, neither the Federal government nor the large corporation plays a major role in incubator formation or management [15]. A senior business advisor at the Massachusetts Small Business Development Center identifies this very point: "Successful incubation is all about connections, and those happen best at the local level. . . . States do their part, but when you look at the best [U.S.] incubators, they break down more by region" [16].

Europe presents a sharply contrasting picture. The European Union cannot match the United States in the number of incubators operating, their robustness of innovation, nor the extent to which they serve to link the academic, business, and financial communities within integrated networks. A recent report on the state of incubators in Western Europe emphasizes that incubators operating in the United Kingdom, Germany, and France are quite modest in size and ambition compared to those in the United States. These incubators tend not to be very entrepreneurial in spirit. In the United Kingdom, there do exist academic incubators in Cambridge, Aston, and Manchester, but adoption by these organizations of innovative ideas is less robust than in the United States. In France, incubators tend to be small and dispersed rather than concentrated in high-tech regions where "bottom-to-top" synergies (agglomeration economies) could take hold. In all three European countries, incubator organization is heavily bureaucratized due to the heavy financial and strategic influence of central public authority (for example, in the United Kingdom, the Ministry of Trade and Industry) and large corporations. This "from-the-top" influence further dampens risk taking and creation of an entrepreneurial mind-set. Table 7.9 shows the discrepancy in number of privately sponsored incubators, especially those created by entrepreneurs, between the United States and the three largest of Europe's economies, individually and collectively [17].

The fact that European incubators do not connect strongly with the marketplace seriously reduces their effectiveness as interlinking agents in the commercialization process. The case of Germany reflects Western Europe as a whole. In Germany, only 12% of companies associated with incubators spring from university-based incubators. In general, the impact of incubators on commercially relevant technology transfer appears to be weak. In Germany, the majority of companies judge the advice given by the incubators as relatively insignificant for them. German incubators and universities have been criticized as being very inefficient and ". . . more focused on fundamental research and . . . they register very few patents . . . and [at best] encourage incremental innovations and improvements . . . rather than radical innovations" [19]. A common concern is that European incubators are not challenged enough and lack an inherent entrepreneurial spirit.

American Corporations, SMEs, start-ups, universities, incubators, state and local governments, and the markets themselves are all part of the complex network—the seamless web—and closely interact with one another to create, diffuse, and adapt new materials technology, and, as a result, propel economic activity. However, in a

Table 7.9. Private incubators (estimates as of October 2000)

	Total number of private incubators (corporate + venture capitalists + entrepreneurs)	Incubators created by etrepreneurs
United States	350	210
United Kingdom	100	60
Germany	60	40
France	30	20

Source: [18].

sense, we have left out a major player in this story, for it is the venture capital community, considered in the broad sense of the term, but distinct from institutional capital, that truly directs and organizes the creation and operation of the seamless web structure. It does so by an intricate architecture of signaling, focusing, and filtering mechanisms that extend throughout, and interact closely with, the seamless web system. The following chapter considers these various mechanisms and the special role played by venture capital in performing this vital function in the advanced-materials technology creation, development, and diffusion network

REFERENCES

1. Table constructed from a number of sources, including Thayer, A. M. (2003), "Nanomaterials," *Chemical & Engineering News,* Vol. 81, No. 35, pp. 15–22; Moskowitz, S. L. (2002), *Critical Advanced Materials Report: Final Draft,* pp. 10–32 (Prepared for: Virginia Center for Innovative Technology, Herndon, VA).
2. See Aftalion, F. (2001), *A History of the International Chemical Industry from the Early Days to 2000,* Philadelphia: Chemical Heritage Press, pp. 397–403 for further discussion on this issue.
3. Refer to Aftalion, F. (2001), *A History of the International Chemical Industry from the Early Days to 2000,* Philadelphia: Chemical Heritage Press, pp. 376-379 for a discussion of how increased regulations, encroaching influence of conservative financial institutions, and other factors has resulted in a once-pioneering industry becoming more cautious when it comes to innovation and technology.
4. Refer to *Nanoinvestor News* (2001–2005), *Company Profile Archives,* to judge the importance of the small start up firms in advanced materials. See also Williamson, O. E., *Markets and Hierarchies: Analysis and Antitrust Implications,* New York: Free Press. Williamson suggests smaller firms are more efficient at developing innovation.
5. According to the information we can glean from *Nanoinvestor News* (2001–2005), *Company Profile Archives,* these figures are probably much too conservative when it comes to advanced-materials firms.
6. Refer to CNI website (www.cni.com) (2003), *Press Releases.*
7. Table constructed from a number of sources, including Moskowitz, S. L. (2002), *Critical Advanced Materials Report: Final Draft,* pp. 5–30 (Prepared for: Virginia Center for Innovative Technology, Herndon, VA); *Nanoinvestor News* (2001–2005), *Company Pro-*

file Archives. Data for earlier years derived from author's doctoral dissertation on the history of petrochemical innovation (Columbia University, New York, New York, 1999).

8. University of Virginia Patent Foundation website (www.uvapf.com) (2002). For further discussion on the role of universities in company R&D, see Ettlie, J. (1998), "R&D and Global Manufacturing Performance," *Management Science,* Vol. 44, No. 1, pp. 1–11.
9. Table constructed from a variety of sources including Moskowitz, S. L. (2002), *Critical Advanced Materials Report: Final Draft,* pp. 4–35 (Prepared for: Virginia Center for Innovative Technology, Herndon, VA); *Nanoinvestor News* (2002–2004), *Company Profile Archives.*
10. Table constructed from a number of sources including; *Nanoinvestor News* (2001–2005), *Company Profile Archives;* Phillippe, A., Bernasconi, M., and Gaynor, L. (2002), *Incubators: The Emergence of a New Industry—A Comparison of the Players and Their Strategies: France, Germany, UK, USA,* Chambre de Commerce et D'Industrie Nice, France.
11. Table constructed from various sources, including VentureChoice website (venturechoice.com) (2004), "Business Incubator: List of Links to Major Business Incubators in the US"; *Nanoinvestor News* (2001–2005), *Company Profile Archives;* New York Biotechnology Association website (nyba.org) (2005), "Business Development: Incubators and Technology Parks (by Region)."
12. Phillippe, A., Bernasconi, M., and Gaynor, L. (2002), *Incubators: The Emergence of a New Industry,* Innoval, p. 9.
13. Ibid., p. 22.
14. Ibid., p. 38.
15. For discussion and excellent examples of this, see Phillippe, A., Bernasconi, M., and Gaynor, L. (2002), *Incubators: The Emergence of a New Industry—A Comparison of the Players and Their Strategies: France, Germany, UK, USA,* Chambre de Commerce et D'Industrie Nice, France, pp. 37–50.
16. Regan, K. (2007), "Nurture Trail: Locally Focused Incubators Thrive in Favorable State Climates," The Journal of New England Technology, May 18.
17. For further discussion of difficulties faced by incubators in the United Kingdom, France, and Germany, refer to Phillippe, A., Bernasconi, M., and Gaynor, L. (2002), *Incubators: The Emergence of a New Industry—A Comparison of the Players and Their Strategies: France, Germany, UK, USA,* Chambre de Commerce et D'Industrie Nice, France, pp. 23–29.
18. Ibid., p. 77.
19. Ibid., p. 45
20. Aftalion, F. (2001), *A History of the International Chemical Industry from the "Early Days" to 2000,* Philadelphia: Chemical Heritage Press.
21. An influential study conducted in the 1970s agued that large multidivisional firms are not efficient in carrying out innovation. They are more successful in the manufacture and distribution of new products to the marketplace. See Williamson, O. E. (1975), *Markets and Hierarchies: Analysis and Antitrust Implications,* Free Press, New York. See also Dougherty, D. (1992), "Interpretive Barriers to Successful Product Innovation in Large Firms," *Organization Science,* Vol. 3, No. 2, pp. 179–202. Dougherty argues that established firms lose innovatve incentives due to the creation of a culture of "routinitation" that is difficult to break down to allow an iconoclastic innovative mindset.

Chapter 8

The Seamless Web II: Signaling, Selection, and Focusing Mechanisms Within the Seamless Network—SMEs, Entrepreneurs, and Venture Capital

Focusing on advanced materials, we have described the teeming network that provides the conduit and connecting points for information and innovation within a technology-driven economy. But we still need to better understand how this network operates to properly select, out of the morass of possibilities, those most promising technologies, and, by the same token, filter out those time-consuming and resource-depleting dead ends. In the previous chapters, we have seen how the United States excels in managing critical risks that threaten advanced materials innovation. Most importantly, through a seamless network, the United States minimizes the risk of a dangerous gap emerging that could create a barrier between upstream ideas or inventions and the practical requirements of the downstream marketplace. The American model of the university–incubator–SME–local government network of decentralized, dispersed technology creation maximizes the opportunity for research-based invention to evolve in response to market-driven values and requirements. This is in contrast to the European approach, which values the large, bureaucratic corporation and centralized "big science" supported by central government. This European model conceives of innovation beginning from the top as a pure idea and filtering down to practice. We know that this scheme imposes significant barriers between even the most brilliant science, a proud European tradition, and the less perfect, hurly-burly world of commerce. This approach falters in practice because, in the European context, the dialogue between researcher, politician, investor, and businessman is uncertain, elliptic, sporadic, and garbled, whereas successful innovation requires this intercourse to be robust, direct, constant, and clear.

But what evolutionary mechanism exists to target in on and pluck out of the vast pool of possibilities just those most potentially fruitful inventions and early-stage conceptions that deserve the special attention of society? Does this happen more readily within the context of the decentralized technology creation network? If so, what sort of heretofore hidden force performs this critically important function by which the few "stars" are identified and nurtured to market? Europe clearly understands the growing importance of this ability of selecting and filtering new technology.

According to the report "Nanotechnology in Europe" (2007), organizations and companies throughout the European Union believe that one of their major shortcomings in competing against the United States is Europe's persistent inability to properly prioritize patents as to their present and future worth [1].

Certainly, the focus and selection mechanisms that apply to the large, integrated firm, which we previously discussed, are severely limited to the confines of that firm and are not usually relevant to the small start-up firm—the real source of innovation in the late twentieth and early twenty-first centuries—nor to the network as a whole. The large firm responds less directly to external market signals, but rather to the endogenous requirements of its internal production and strategies. It pursues advanced materials as in-house intermediates that augment its product line and allow it to respond to the shifting demand climate for its final products and technologies. Here, advanced materials serve as a critical part of a firm's internal, self-enclosed value chain.

The small start-up is not yet well enough developed as an organization to commercialize new materials as intermediates for its own operations. The development and making of a new material—nanotubes, conducting polymers, and so forth—often is the sole activity of the firm.

The start-up firm (i.e., the SME) then must look to the outside market as its guide to what innovative avenue it should proceed down. By definition, such SMEs do not have sufficient history or resources to claim any permanent, on-going R&D program. Unlike many large firms, they do not have the option of pursuing different development routes. Their success depends on efficiently—in terms of time and resources—commercializing a particular advanced-material technology that it licenses, often from universities and government laboratories. What it can focus on and select—where it has choices—is the final form that the technology (i.e., its processes and products) takes within the general orbit of the original patent description and license agreement. For the small start-up firm, advanced materials are part of a value chain that is not self-contained within a single firm but winds its way between firms and even industries.

SELECTION AND FOCUSING MECHANISMS OF START-UP ADVANCED-MATERIALS SMEs

Unique types of mechanisms focus the smaller advanced-material firm on, and compel it to select, certain technology avenues over others. In the case of the large corporation, as we have seen, internal signals guide innovation, but in the case of SMEs, market signals from outside predominate. There are three main models that apply with respect to these advanced-material firms. These models help us at the least to get an approximate sense of the funneling and filtering processes that emerging innovative SMEs embrace in the pursuit of capturing the ripest and sweetest fruit from society's existing and multifarious technology tree.

The Upstream Cost Savings (UCS) Model

Price reductions and quality-improvement mechanisms are critical market tools for the small start-up firm in determining the type of advanced-material technology to adopt. Specifically, start-ups must, for this sort of market, be vigilant in the production process with which they make their materials. These are not necessarily as important for the larger firm that pursues advanced materials as intermediates for its own products and processes. For the small start-up, the question here is how best that is, most economically, to produce new types of materials in sufficient volume, at low enough costs, and with sufficient purity (or other characteristic) to compete with other materials as intermediates for the outside original equipment manufacturing (OEM) market.

The petroleum refining and petrochemicals industries are examples of the type of market that compels advanced material suppliers to adopt the UCS model. In general, these industries place a premium on the cost of materials and intermediates. They are volume, mass production producers who process enormous amounts of inflows and outflows of materials daily. These industries already have minimized the costs of labor through years of improvements in mass production technology. It is the cost of their material inputs that directly impact already small per-unit profit margins.

Carbon Nanotechnologies Inc. (CNI) faced this issue when it licensed a technology to make single-wall carbon nanotubes. Although the original patent offered a general blueprint for a viable process, the exact technology depended on the company's own markets and their requirements. In particular, clients in the oil industry looked to CNI to supply it with nanotubes for use in catalytic processes. The stumbling block was the cost to the refiners, who deal in volume throughput and have narrow profit margins. CNI found a route that offered a way to significantly and uniquely cut the cost of making nanotubes. The process developed by CNI, the so-called HiPco process, modified the original patent by using a by-product of the refining industry, carbon monoxide, under high pressure, thus creating critical economies of scale. To further streamline and tailor the design of the technology, CNI collaborated with the chemical and petroleum engineering contracting firm Kellogg, which in years past worked with the refining and petrochemicals industry in developing important processes such as ethylene cracking, fluid catalytic cracking, and ammonia production, to scale up the nanotube reactor and to seamlessly integrate it into the petroleum refining plant.

The Downstream Cost Savings (DCS) Model

For some clients, advanced materials promise to reduce the costs of making their products or performing their services. In such instances, considerations for OEMs in the cost of labor, capital, or energy can override considerations of the costs attached to the advanced-material inputs themselves.

We note that in a number of market applications only a very small amount of an advanced material is required to impart desired properties to a traditional, low-cost polymer. In such cases, the OEM making that heightened hybrid material and parts from it is less concerned (within certain limits) about the cost of the advanced-material additive. For instance, material costs may play a minimal role in the plastics extrusion and molding industry.

Let us again turn to Carbon Nanotechnologies Inc. (CNI) for a more specific illustration. In addition to the petroleum industry, CNI develops carbon nanotubes for polymer processors, including extruders and casters. Although the price of the nanotubes is certainly of importance to this industry, CNI's materials could not succeed in the polymer processing industry if they did not make the extrusion or casting processes themselves easier and cheaper than currently exists. As does the petroleum industry, this industry also operates under very tight per-unit profit margins. But this is a more labor-intensive industry, relying as much on craft and expertise as on new technology. In the industry, the margins shadow closely the speed and costs of extrusion. Any new material that causes disruptions to the production flow, or requires extensive retraining on the part of labor, would not be selected to replace the more traditional and familiar plastic materials.

Faced with this situation, CNI took a different route than it did with the petroleum industry in responding to these particular market signals. Remaining within the purview of the fundamental patents, CNI focused on a new material to reduce production costs for its clients in the plastic processing industry. It researched, developed, and introduced a new form of single-walled carbon nanotubes called "BuckeyPearls." CNI designed this material to have the specific properties of carbon nanotubes, yet be more easily processed in extruders and other polymer-processing equipment.

The Downstream Product Enhancement (DPE) Model

A new material technology not only augments a client's process, but it may play an even more important role in enhancing his product. This, rather than cost savings and process enhancement, tends to be more important in high-technology markets, including electronics, biotechnology, pharmaceuticals, and aerospace. We note here that the existence of close technological and economic complementarities between the material and the OEM product is most important. We have seen that large firms tend to focus on new materials that are complementary with, and thus will enhance, that firm's internal products. A similar situation applies here, except that now the small start-up tailors his materials for an outside OEM client.

Within the biomedical field, pharmaceutical companies, in response to market demand, look to the innovative advanced-materials firm to provide added benefits above and beyond the action of the drug agent itself that they themselves cannot satisfy. Advances in drug delivery technology increasingly require the use of new material designs. For example, new types of polymers, nanoclays, and biochemical materials, coupled with innovative encapsulation techniques, are today creating a revolution in drug delivery technology that makes it easier to regulate the amount and timing of drugs released in the body.

These materials even allow previously ineffective agents to take on new life due to improved drug delivery capabilities. So we find that the advanced-material firm NanoSystems responded to the market needs of its major client, the European company Elan, which discovers, develops, manufactures, and markets therapeutic products in neurology, pain management, and autoimmune diseases. Nanosystems developed nanocrystal technology that renders previously insoluble biochemicals capable of entering solution in the body by transforming them into nanoparticles. This technology permits previously rejected drugs to be placed on the market as pharmaceutical agents in common dosage forms that can be taken by mouth or injected.

The examples that illustrate the above three models help to direct us to which types of OEM industries are more likely to impose particular sorts of focusing and selection forces on the advanced-materials sector. The matrix in Table 8.1 displays this distribution.

As useful as these models of new-technology capture can be, they do not provide us with the Rosetta Stone that unlocks the inner secrets of how the innovative U.S. firm has become so spectacularly precise and efficient in identifying out of all apparently interesting candidates just those early inventions (usually in the form of patents) that will most likely win commercial success. After all, in any one of the models described above, an SME, or even a larger corporation looking to augment its stagnating R&D capability through licensing or merger and acquisition of a smaller technology-driven firm, will be faced with a wide spectrum of possible routes, one as likely to lead to byways and dead ends as to success. The various internal (if a large corporation) or external (if a SME) pressures do not dictate which of the many possibilities screaming for attention ought to be favored in order to relieve those pressures. How, then, in this welter of possibilities, does one choose wisely so as not to waste precious time and resources on will-of-the-wisps and so expose itself to intense competitive incursions from hungry and resourceful foreign companies just chafing at the bit to stampede across the global marketplace ahead of American enterprise? The following sections address this issue.

Table 8.1. The small start-up firm—types of OEM industries associated with the different selection and focusing mechanisms

	High-technology industries	Commodity-processing industries
UCS Model		X
DCS Model		X
DPE Model	X	

SMEs AND THE VENTURE CAPITAL COMMUNITY

Generally speaking, efficient financial markets are most critical for technological development and economic progress. This is especially true in the late twentieth and early twenty-first centuries when, unlike large corporations, SMEs depend mightily on outside funding during the critical early and middle stages of their life cycle.

Significant differences in patterns of private investment activity distinguish the United States and European Union in terms of their capability in transferring the results of research and development into the marketplace. It is clear, even to Europeans, that the heart of new technology is venture capital and that this sector within the European countries is not as vibrant or wide reaching as it is in the United States. So we find that in its 1999 report on Biotechnology clusters, the U.K. Ministry of science observes that "The U.S. venture capital industry is the most mature in the world and it has been a major contributor of [advanced] U.S. biotechnology" [2]. This is not the case at all in the European Union, where venture capital remains in a relatively nascent state. Despite the fact that officials in the European Union consistently push the development of a robust European venture capital industry as a driver of future innovation and growth, they admit that they still have a long way to go to match U.S. achievements in this arena. Moreover, there is evidence that Europe is looking to the U.S. experience as a model for structuring their own venture capital industry [3].

In fact, there is a significant gap in performance between U.S. venture capital firms and their European counterparts, especially in terms of effectiveness, as measured by rate of return. As it is, European venture capital geared toward innovation and early stage financing only really took off in Europe in the late 1990s. Even so, the aggregate investment by European venture capitalists reached 12 billion Euros in 1999, compared to $105 billion for U.S. venture capital in 2000. Table 8.2 shows the rise of American venture capital placements. In the year 2000 alone, the U.S. venture capital industry financed about 6000 firms. The median amount invested per round of financing increased from $5 million in 1998 to $11 million in 2000.

The issue here is not only about amount of venture capital invested, but where and when (in a firm's life cycle) that investment takes place. The United Kingdom unquestionably has the largest and most active venture capital community in Europe; and we see U.S. venture capital placement retrenching after 2000 in response to a general economic slowdown. However, the U.K.'s venture capitalists invest precious little money in start-up companies, where, as we know, most twenty-first century innovation takes place. In 2000, the total venture capital investment was 8 billion Euros (compared to 4.8 billion for Germany and 5.3 for France), but only 1 billion Euros of this money found its way to start-ups. In this same year, the U.S. venture capital community invested in initial stages of companies (seed + start-up) seven times more money than the United Kingdom, France, and Germany combined, and nearly four times more than Europe as a whole. Further, the United States invests more intensely in these start-up firms than the European venture capitalists spend on their companies. Whereas in 2000 the United States invested in

Table 8.2. Venture capital trends in the United States ($ billions)

	Venture capital financing, total
1991	1.9
1992	5.1
1993	3.8
1994	7.7
1995	10.0
1996	11.2
1997	14.9
1998	21.2
1999	54.2
2000	105.2
2001	40.7
2002	21.8
2003	19.7
2004	22.1
2005	22.8
2006	26.0

Source: [4].

half the number of companies (2299) than did Europe as a whole (4676), on average the Americans invested eight times more per company than did the Europeans. These numbers strongly suggest that, compared to the investors in the European Union, the U.S. venture capital industry is far more selective of the companies that it takes on, devotes more money to the most promising companies that it selects, and is far more successful in targeting and nurturing the best candidates for growth [5].

Fundamentally, The U.S. venture capital industry differs significantly from its counterpart in other countries in two significant ways. First, it has developed into a more specialized industry that can economically shepherd through the maze of the commercialization process the promising technologies that otherwise might "fall through the cracks." Second, it stands at and controls the very center of the seamless web system from which productivity-enhancing innovation emanates. From that pivotal position, it integrates the U.S. capital, market, and science–technology communities into the closely interconnected information network that is essential for technological development and its end result, economic growth.

Specialization in the U.S. Venture Capital Community

More than the mere presence of investment capital, industrial competitiveness depends on a varied and seasoned venture capital community. Generally speaking, when first starting up, advanced-material firms rely on the financial support of savings, families, and friends, as well as on government grants. This funding "gets things going" and pushes companies and research projects in certain directions, but this phase often is one reflecting enthusiasm for new ideas and designs rather than a finely tuned sense of what demand exists, much less where it may thrive. At this point then, the market does not yet come into sharp focus. The actual linking of the firms to the real-world marketplace typically occurs later on, under the direction of the venture capitalist, who begins to sort out the most commercially promising companies and technologies. However, the range of technologies enters different markets at various points in their development. As a consequence, young, high-growth firms benefit significantly from being able to tap into a continuum of complementary capital sources offering specialized finance options that focus on the needs of particular industries and on specific phases in the product (and firm) life cycle.

A distinctive element in the U.S. venture capital structure is the sheer variety in the types of venture capital available that can push firms toward specific market arenas at certain and appropriate points in their life cycles. This concentration of talent and resources generates significant value through economies of specialization [6]. This diversity allows investors to specialize in the types of firms serviced. Venture capital firms like Harris and Harris focus on nanotechnology, which includes nanomaterials, thin-film materials, and so forth. Economies of specialization over time builds into a venture capital group an innate ability to sniff out and uncover, from the morass of possibilities, up-and-coming technologies and to guide these to commercial fruition. Specialization in venture funding of advanced nanomaterial development is particularly well developed in the United States. Approximately $386 million in American venture capital was invested in domestic nanotech and nanomaterial companies in 2002. Overall, about $900 million in venture capital went to nanotech and nanomaterial companies between 1999 and 2002, approximately six times greater than the case for the European Union. Companies within the European Union find venture funding of such technologies less than forthcoming as, according to that U.K. Biotechnology Cluster Report, "companies [in the United Kingdom] are experiencing difficulties securing funding for continued growth" [7].

In addition to U.S. venture firms specializing in nanotechnology and advanced materials, they tend to concentrate in servicing firms at a certain point in their life cycles. Since the late 1980s, U.S. venture capitalists have been dividing themselves into "high-end" and "low-end" segments. In the former case, they handle only firms (the later-stage companies) that have reached a minimum revenue level. These high-end venture capitalists understand and cater to the technical, managerial, and marketing needs of firms that have reached a certain degree of economies of scale. On the other hand, they avoid the smaller, early stage start-ups. Yet, these start-ups are just the ones who energize the economy with truly innovative products. Whereas overseas, these engines of economic growth often fall through the cracks, the United States offers a particularized venture capital segment—the angel investor—that keeps start-up firms financially (and otherwise) afloat until they can stand on their own commercially.

The concept of the angel (or "informal") investor is rooted deeply in American culture. (The term itself was coined in New York, referring to people who invest in Broadway shows.) Angels band together to pool and better focus resources. They are useful in getting a company started. They come in earlier than other financial players and make smaller investments. Increasingly, angels band together to form investment groups with as many as 100 members or more with average infusions per group anywhere from $25,000 to $2 million. In addition to doctors, real estate developers, and well-heeled relatives, they draw from the pool of individual entrepreneurs and technology experts, CEOs, lawyers, scientists, and consultants. These investors grew rich in the 1990s from rising stock prices and the continuing wave of initial public offerings and company acquisitions. As a group, they control a broad network and deep body of experience. They remain active in different industries, which they tend to know very well. They also know the trends. They know what will work. Many serve as executives or board members of start-ups that the group funds. They understand the latest technology and its time frame, and are quite willing to back an early stage idea to the point when a later-stage, "high-end" venture capitalist can take over. Observers of angel investing emphasize the fact that angels are very selective in what they back. They look most favorably on companies that they believe will reach carefully laid-out benchmarks of success. They take such projects to their first stages, thus giving them time to mature before interacting with later-stage investors [8]. The U.S. Small Business Administration estimates there are 250,000 angels in the United States committing $20 billion annually to over 30,000 private companies. Percentage-wise, the United States displays a great intensity of angel or informal investment. In the United States, in 2000, 150 million adults (18–64 years of age) invested $196 billion in angel capital. This amounts on average to $800 per adult investor compared with nonangel venture capital in United States of $600 per adult. Thus, on average, every dollar of venture capital is matched by $1.6 in angel investment, an important measure of commitment to the most innovative start-up firms.

As with venture capital in general, Europe has fallen behind the United States in its angel business activity. Although angel investment appears to be on the rise in Europe (as demonstrated with the growth in Europe of national networks and associations such as EBAN and improvements in tax regimes favorable to angel investors), EU's angel community has not yet achieved the size and status of its American counterpart. Angel money in Europe is not as abundant as in the United States nor is it applied as intensely. As we see Table 8.3, none of the three major Western European countries can match the United States in terms of proportion of adults active as angel investors.

Table 8.3. Informal (angel) investments (2000)

Investment per adult (18–64 years) (US$)	United States	United Kingdom	Germany	France
Venture capital	600	100	50	—
Informal (angel) capital	800	490	110	—
Total	1,400	590	160	—
Proportion of adults (18–64 years) who invested in a company start-up	5.3%	2.6%	3.2%	1.6%

Source: [9].

If we compare angel investment per adult, the three European countries combined do not come close to the level of investment reached by the United States. This can only mean that European start-ups must find it more difficult to obtain the money they need for research and development and for moving into the commercial phase. Compared to the United States, Europe does not sustain a high level of capital placement in early (angels) and later (venture proper) development. The United States leads all the OECD countries in terms of early and expansion-stage placements as a share of GDP. In the period 1998 to 2001, this figure for the United States was 0.55% compared to only 0.18% for Germany, 0.22% for France, and 0.3% for the European Union as a whole [10]. We can say then that the United States outpaces Europe (and the rest of the world) in the depth and breadth of its investment pool that can specialize in the development and commercialization phases of new technology development. It is clear that the United States is far better positioned to support its most advanced start-up firms and capture the lead in the newest technologies.

The U.S. venture capital industry excels as well in its geographical reach. States and localities within the United States actively seek out and nurture grass-roots venture capital activity in order to attract high-technology firms within their areas that can generate income, jobs, and economic growth. To this end, several states have created venture capital funds to nurture the development of local businesses. United States venture investment tends to be concentrated in a few key regions, such as Silicon Valley (CA) and the Boston area (MA), two areas leading in high-tech venture capital activity between 1980 and 2000. In late 1990s, other states and regions began to grow as centers of venture capital, including Oklahoma, Louisiana, New Hampshire, Pennsylvania, North Carolina, New York, Maryland, Colorado, and Texas.

The Capital, Business, and Science–Technology Network: The "Dialogue" Factor

Variety, specialization, and geographical reach are not the whole story of American venture capital leadership. Venture capital, broadly defined, effectively facilitates dialogue between the science–technology and the business–market communities, an interchange that is vital to maintaining the rate and relevance of technological change. In this sense, venture capitalists often act as the critical gatekeepers, spanning and policing information flows between different professional arenas and selecting only the most promising for further development. This role of the venture capitalist in assessing the market and then selecting and guiding high-potential candidates for that market is critical and, arguably, of even greater importance than as a source of funding.

In the United States, to a greater degree than in other countries, dialogue and information flows easily between the venture capital community and the scientific, engineering, and business networks. Within the United States, such networking is one of the most important methods of linking capital with R&D for commercial advantage. In interviews of advanced materials managers and engineers conducted by the author and his team at conferences within the U.S. and Europe indicate that, approximately half the time, deals emerge out of personal relationships between entrepreneurs, a number of whom have technical backgrounds, and scientists and engineers. A significant number also come from referrals from technical and business people the venture capitalist knows and trusts. Conferences bring together venture capitalists and management of advanced-materials start-ups (Table 8.4).

Europe is not the only region that suffers in comparison to the United States in creative dialogue between the technical, financial, market, and political realms. Certainly, other countries have wealthy enough investors interested in getting involved in promising projects, but this in itself does not guarantee technology growth. Israel is an interesting case and one that reflects the same set of problems in technology creation as the European Union. Israel has one of the largest venture capital markets outside the United States. With over $3 billion invested annually by the sector, it is, according to Price Waterhouse Coopers, the sixth largest venture capital market in the world. However, despite a growing and creative research and engineering pool, its venture capital community has fallen behind other industrialized countries in entering into advanced materials, nanotechnology, and other high-technology fields. There are currently about 30 Israeli companies operating in these areas, none as yet a major player internationally. The major problem in Israel appears a lack of communication between the venture community and the advanced-material and nanotechnology industries. In Israel, as in the European Union, scientists and engineers who attempt to commercialize their discoveries find themselves "talking over the heads" of potential investors. At recent conferences held in Israel arranged to bring together scientists and the business community, John Medved, Managing Partner of the Jersusalem-based Israel Seed Partners, reported that scientists simply read from their recent and highly technical papers [12]. Since most venture capitalists in Israel, the European Union, and other countries do not have technical expertise in this area, they have trouble seeing how to transform these seemingly abstract concepts into commercial products. This lack of communication points to a serious divide between the research establishment and the financial and business communities.

Recent investigations by European organizations support the conclusion that, like Israel, the European venture capital community has been unable to bridge the communication gap with the real-world market. These studies conclude that the U.S. venture capital community had sharper screening skills than Europe's. This means that, in the United States, there is a higher degree of translating initial investments and funding into commercial success. European observers note that U.S. ven-

Table 8.4. Percentage of advanced-material start-up firms funded by venture capital that established initial contact at conferences

Region	1980	1990	2000	2005
United States	10	17	28	35
European Union	2	5	11	19

Source: [11].

ture capitalists excel at determining whether a candidate technology has a realistic time and cost schedule, whether a sufficient market exists and is on the upswing, and if competitors, especially with proprietary technology, could pose a potential threat. To further strengthen the linkage between company, technology, and market, venture capitalists within the United States take a proactive role in the fledgling company such as purchasing stock, helping set up the corporate structure, working with the company to put together the management team, and even serving on the new company's Board of Directors.

This is decidedly not the case in Europe (or other countries, such as Israel). As an EU report observes, ". . . venture capital firms in Europe are more deal makers and less active monitors; they seem to be still lagging [the United States] in their capacity to select projects and add value to innovative firms" [13]. The European Commission concludes that "capital is plentiful in the European Union" but the European investment community is neither as accurate nor as swift as its U.S. counterpart in identifying the early and most vital entrants in the technology race and doing so in a timely manner, that is, before the market opportunity passes by [14].

The issue for European venture capital is its penchant for investing in well-known industries, as opposed to new, emerging technology. Whereas U.S. venture capital money tends toward the very latest high technology enterprise, especially new materials and their consuming industries (IT, biotech, and energy), the greater proportion of European venture capital goes to those areas that the Europeans have long held a reputation for mastery in, including industrial machinery and equipment, fashion, and leisure products (sporting goods). About one-third of all E.U. venture funds find their way to these industries compared to less than 10% in the US [15].

This explains the conclusion of Ramon Compano, member of the European Commission's Program for Future and Emerging Technologies, that European advanced-material and nanotech ventures receive far more support from public investment, which has little understanding of the salient characteristics of the marketplace, and research grants than from the private sector. Table 8.5 shows trends

Table 8.5. Advanced materials and venture capital (%), United States versus the European Union

Year	European Union	United States
1970	2	20
1980	5	35
1990	10	50
2000	18	65
2010	27	75
2020	38	80
2030	49	85

Source: [16].

in private venture capital in advanced-materials R&D. It displays the percentage of R&D investments supported by the investment community, both within the United States and the European Union. The table confirms the generally preeminent role venture capital plays in the United States and also the growing importance of venture capital in U.S. advanced-materials R&D relative to the European countries.

In the previous chapter, we focused in on the complex warren of corporations, SMEs, universities, and incubators that compose the seamless-web model of advanced-materials growth. In this chapter, we explored still another layer of activity that ties together and integrates these various actors within the networks. Selecting and focusing on mechanisms within the management structure of SMEs and, even more importantly, as carried out by the venture capital community, helps to strongly link the various actors in the creative network and create a self-contained, highly integrated technology "zone."

These discussions lead us directly to the idea that, far more than in the European countries, the advanced-materials industry in the United States alienates itself from the top-heavy, dictatorial, and ultimately ineffective "big-science" complex that imposes from above an aristocratic theoretical imperative onto an unsuspecting and too often unreceptive marketplace. The latter generally does not take too kindly to unsolicited directives from above and emanating from the alliance of big government and the scientific community that too often isolates itself from the needs and workings of the real world. In place of this top-heavy, "control from above" model of economic growth, we posit a type of "bottom-up" structure found in the United States, a complex, pervasive and intricately interconnected network of organizations, institutions, and government agencies (state and local) that establish strategic pathways linking empirical research and practical engineering design activity taking place throughout society to the marketplace. These pathways capture and facilitate the back and forth flow of information, skills, and signals from one sector to the other. Research and development proceeds iteratively and only with close and constant communication with the ultimate users as to their perceptions and desires. The research and development establishment, such as the university materials engineering department and the high-technology start-up firm, proceeds in accordance with the requirements of empirically based engineering design, but regularly adjusts the content and form of products to signals received from the "outside" world that defines the demand curve.

Through this delicate dialogue between "technology push" and "demand pull," a balance is achieved through continual communication between those creating and those employing what will be created, and a hybrid model of productivity-inducing technology creation evolves. This pervasive, interconnected network is by no means created willy-nilly within society, in an inchoate and unpredictable manner. Creation from below does not mean chaos and disorder. Nor is the seamless web of creation monolithic, but a number of web structures located in particular geographical areas, with (more or less) well-defined borders, and organized in particular and rational ways. The following chapters consider the dynamics of the process that generates these critical centers of local and national growth, where and how such webs of creation come about, and the laws that control their evolutionary paths.

REFERENCES

1. Nanoforum (2007) *Nanotechnology in Europe—Ensuring the EU Competes Effectively on the World Stage,* Survey and Workshop, Final Report (Dusseldorf, Germany), p. 20.
2. UK Minister for Science (1999), *Biotechnology Clusters,* Government Report, Ministry of Science, London, England, UK: p. 34.
3. See Baygan, G. (2003), Venture Capital Policy Review: United States Directorate for Science, Technology and Industry (STI) Working Paper 2003/12 (OECD: Paris, France), p. 5. See also Nanoforum (2007) *Nanotechnology in Europe—Ensuring the EU Competes Effectively on the World Stage,* Survey and Workshop, Final Report (Dusseldorf, Germany), p. 54.
4. National Science Board (2008), *Science and Engineering Indicators 2008—Volume 2,* National Science Foundation (NSF), Arlington, VA, Appendix Table 6-57.
5. See Hege, U., Palomino, F., and Schwienbacher, A. (2003), *Determinants of Venture Capital Performance: Europe and the United States,* Risk Capital and the Financing of European Innovative Firms (RICAFE) Project (Financial Markets Group: London, UK), p. 4, which confirms that ". . . US VCs have sharper screening skills than their European counterparts. This translates into a larger fraction of the total investment invested in the initial round and a higher degree of translating initial investments . . . into success."
6. See Baygan, G. (2003), *Venture Capital Policy Review: United States,* pp. 6–7. See also Hege, U., Palomino, F., and Schwienbacher, A. (2003), *Determinants of Venture Capital Performance: Europe and the United States,* p. 6–7, which refers to U.S. venture capital as having a distinctive ". . . specialization [that] clearly may be a source of value creation, as VCs presumably are more expert in the stage specific skills of their contribution." Specialization can be in phase (or stage) of business or in type of technology. For example of the latter see Mason, J. (2002), "VC Firm Harris & Harris Specializes in Small Tech," *Small Times,* February 26.
7. UK Minister for Science (1999), *Biotechnology Clusters,* Government Report, Ministry of Science, London, England, UK: p. 26.
8. For discussions of US business angels and vis-à-vis the European situation, See Baygan, G. (2003), pp. 18–19 and *Foreign Direct Investment Magazine* (www.fdimagzine.com); Thuermer, K. (2003), "Small is Bountiful," Foreign Direct Investment Magazine online (www.fdimagazine.com, London, UK; Financial Times Group): August 5: pp. 1–5; see also Nanoforum (2007) *Nanotechnology in Europe—Ensuring the EU Competes Effectively on the World Stage,* Survey and Workshop, Final Report (Dusseldorf, Germany), p. 20.
9. Phillippe, A., Bernasconi, M., and Gaynor, L. (2002), *Incubators: The Emergence of a New Industry: A Comparison of the Players and Their Strategies, France–Germany–UK–USA,* Research Report, Nice, France; CERAM Sophia Antipolis and DiGITIP (French Ministry of Economy, Finance, and Industry), p. 71.
10. Baygan, G. (2003), *Venture Capital: Trends and Policy Recommendations,* STI Working Paper (OECD: Paris, France), p. 7.
11. Interviews used in constructing Table 8.4 were conducted both in person and via phone

and e-mails between 2005 and 2007. A standardized interview questionnaire was employed for each interview. Interviews were conducted from contacts made at conferences held in the United States and in Europe. From all interviewees, 40% were managers (almost all with technical backgrounds) and 60% scientists and engineers. Interviewees worked across the major advanced material groups and their allied industries.

12. Machlis, A. (2001), "Israel's Big Science, VC Sectors Are Just Waking up to Nanotech," *Small Times,* December 10.
13. Hege, U., Palomino, F., and Schwienbacher, A. (2003), *Determinants of Venture Capital Performance: Europe and the United States,* p. 4.
14. Commission of the European Communities (1999), *Towards a European Research Area,* Communications from the Commission (Brussels, Belgium; European Union Publication), p. 1. This has been a constant concern for the Europeans, at least since the 1990s. For example, see Commission of the European Communities (1999), *Towards a European Research Area,* Section, "More Dynamic Private Investment: Encouragement of the Creation of Companies and Risk Capital Investment."
15. Rausch, L. M., *Venture Capital Investment Trends in the United States and Europe* (1998), Issue Brief, Division of Science Resources Studies, (Arlington, Virginia: National Science Foundation), p. 4.
16. Table constructed from data and information contained in a number of sources, including: Baygan, G. (2003), *Venture Capital: Trends and Policy Recommendations,* STI Working Paper (OECD: Paris, France), pp. 6–17; Hege, U., Palomino, F., and Schwienbacher, A. (2003), *Determinants of Venture Capital Performance: Europe and the United States,* Risk Capital and the Financing of European Innovative Firms (RICAFE) Project (Financial Markets Group: London, UK); Rausch, L. M., *Venture Capital Investment Trends in the United States and Europe* (1998), Issue Brief, Division of Science Resources Studies, (Arlington, Virginia: National Science Foundation); Moskowitz, S. L. (2002), *Critical Advanced Materials Report: Final Draft,* pp. 4–35 (Prepared for: Virginia Center for Innovative Technology, Herndon, VA).

Part Five

Organizing Growth: Clusters and Gatekeepers

Chapter 9

Clustering and Synergies

An examination of the dynamics by which the seamless web evolves takes our exploration of the advanced-materials industry to a consideration of the formation of high-technology clusters within specific geographic regions. This discussion touches closely on an understanding of how strategic management of this technological process must be carried out in a directed and coherent manner within a multiple-environment context.

INTRODUCTION TO ADVANCED-MATERIAL CLUSTERS [1]

The heart of advanced-materials creation is in the cluster. Success in developing industrial clusters that produce new materials distinguishes the United States from other countries. Potential investment in advanced materials depends on where and when clusters will occur. Rarely can advanced-materials technology form outside of the synergies inherent in the cluster environment, and companies that develop new-material technologies look for clusters in which to attach themselves as this provides them with the greatest potential for growth.

Clusters are geographically localized groupings of firms with backward and forward linkages, tied together by commonalities and complementarities. Geographical concentration is important in order to maintain the intense interchange needed. The agglomeration of producers, customers, and competitors, based on geographical proximity and linked by complementary expertise, promotes positive externalities through increased innovation, and especially through technology spillovers, and maximizes efficiencies and increased specialization. Clusters benefit companies in a number of ways, including the action of economies of scale, deeper and more specialized (and cheaper) labor and capital markets, and information and resource transfer effects that assist and even spearhead the innovation process. For example, studies of the relationship between social capital and knowledge acquisition found that social interaction—such as occurring in clusters—facilitates learning by "fostering close, intensive information exchange" [21].

In studying advanced-material clusters, we must distinguish between "mature" and nonmature or new clusters. Most research involving European clusters (such as in Italy, Belgium, and France) relates to mature clusters involving large firms in traditional industries. Such clusters tend to be specialized in that they are organized around similar products, processes, or services. For example, there are clusters in automotive production, ceramic products, steel, industrial chemicals, and so forth.

On the other hand, clusters can be a concentration of actors and organizations within highly innovative and emerging industries or sectors. These "new" clusters rely on the transfer of information, skills, knowledge, and technologies between firms. This process depends more often than not on people transfer within and between firms. Recent research on organizational knowledge has emphasized the importance of tacit and location-specific knowledge. This research emphasizes the importance of "social capital" that is embedded in historically accumulated reputations, systems of trust, and social networks. Firm-level productivity often depends on historically developed informal social networks that extend beyond firm boundaries.

The firms themselves in such clusters are often comprised of small- and medium-sized companies, including many start-ups, that depend on venture money rather than institutional capital for growth. The importance of the modern cluster for nurturing and supporting innovative SMEs cannot be overemphasized. SMEs, which as we have seen tend to be more innovative than larger corporations, are also less integrated, which can limit their ability to commercialize their innovations. Within a cluster environment, they do not have to bear the entire burden of developing new technologies, finding new markets, training skilled workers, or raising capital. Many of the costs of specialization are shared by or embedded in a dense network of institutions. Entrepreneurial SMEs benefit because the linkages allow firms to enhance their strategic marketing options and to compete effectively in, or circumvent, channels normally controlled by larger firms, which serves as a source of competitive advantage. In short, the cluster structure is better able to succeed than companies and industries acting separately without the benefit of resources that complement one another.

The dynamic of cluster formation is particularly noticeable in the case of advanced materials. In certain regions, we find that suppliers, as well as more downstream processors, enter into such cluster relationships, and for the same reasons. Advanced-materials clusters continue to innovate as new and different companies enter into the cluster environment. Such a company, for instance, often acts as a source of demand for a new material, possibly a purer type of nanotube, which is then researched and developed by one of the advanced-materials firms in the cluster. The close proximity of these firms in the group results in synergies that are often crucial to accelerating this development.

ADVANCED-MATERIAL CLUSTERS IN THE UNITED STATES

In previous periods, high-technology clusters that evolved within the United States tended to concentrate in that region which excelled in one resource or another. The Niagara Falls region of New York State gave rise to the first important industrial cluster of the twentieth century. Cheap power provided the central resource, in this case in the production of important chemicals and metals such as aluminum. In the

1920s, the first petrochemical cluster arose in the Kanawha Valley of West Virginia, which was rich in natural gas deposits. Following World War II, the Southwest, with its extensive oil reserves, saw the emergence of the largest petrochemical cluster in history. This region supplanted the Kanawha Valley as the center of energy and materials production. In the1970s, electronics and telecommunications pushed petrochemicals aside as the high-growth technology. As it did so, the central technological cluster shifted once again, this time from the Southwest to the West Coast and Silicon Valley. In this case, intellectual input, rather than raw material supply, anchored this revolutionary technology cluster.

By the 1980s, the new high-technology clusters once again formed around advanced-materials firms, often as needed by the computer and telecommunications industries, as well as biotechnology and other industrial activity. By that time, however, a more democratic trend became evident. Rather than one region dominating the others due to an authoritarian control over one critical resource or factor of production, many regions began developing simultaneously their own unique and competitive advanced-material groupings.

In each of these cases, we discern certain common patterns. There is the pivotal role of intellectual property, especially as developed within university engineering departments and managed by university technology transfer offices. On the other hand, the Federal government, through its funds and other resources, exerts only a minimal influence here. Typically, advanced-material clusters in the twenty-first century United States evolve internally from other companies, the resources provided by insider venture capital, and local governments with a direct stake in the job formation and economic growth that would emerge from concentrated innovative efforts in their own backyard. In other words, the most robust advanced technology clusters develop at the regional and even grass roots level [2].

Such creative cluster activity operates in complex or multidimensional environments. That is to say, they encompass and closely interrelate a wide range of realms and disciplines—economic, market, scientific, technical, financial, political—as well as industries and firms. Such a creative cauldron containing this diversity is necessary because true innovation requires that all these entities—the actors and organizations associated with them—must transfer knowledge, information, capital, technology, and so forth continuously and often simultaneously across multiple boundaries. Examination of the most successful clusters clearly tells us that competitiveness and innovation require agglomeration of highly interrelated industries that are linked through closely interacting economic, market, technical, geographical, and political forces [3].

In general, too great a focus on one component, say the scientific or the political, over others creates imbalances in the innovative process and ultimately causes bottlenecks and a scaling down of the R&D effort, and thus reduced efficiency and effectiveness of the final technology. It may even halt the innovation process altogether, allowing other competitor companies and clusters to gain the upper hand in the market. Indeed, recent studies show the dynamic entrepreneurial and competitive culture in U.S. high-technology clusters, such as those along Route 128 (Massachusetts) and within Silicon Valley (California). Behind such clusters, we observe a complex and closely integrated multicomponent culture promoting growth, innovation, and economic development. In these areas, an infrastructure to support entrepreneurship existed as well as capital (venture), market research, knowledge, and the drive to commercialize innovations. Increasingly, the small to medium sized entrepreneurial firm operates within—and indeed helps to form—innovative cluster environments [22].

The case studies that follow illustrate the wide variety of advanced-material clusters that are emerging within the United States. These examples show their wide and expanding geographical reach, representing the major regions of the United States. They also illustrate the rich variety and dynamism of the "cluster culture" within the U.S. advanced-materials industry. Each example represents a different model for advanced-material cluster formation. In certain cases, clusters form around dominant companies; in others, regional political influence holds sway; and in still other cases we find clusters based on the revitalization of older industries or preexisting clusters shaking out the cobwebs and renewing themselves as dynamic agents for advanced twenty-first century technology. Nevertheless, even though each regional cluster forms and evolves under unique circumstances and under different evolutionary forces, in all cases the fundamental concept of these intercompany structures emerging from complex or multicomponent environments remains in force. In all cases as well, even those that began through the auspices of large corporations, show the growing role of the small, innovative firm as cluster leaders. These examples from major or promising advanced-materials companies operating within, and acting as flagship enterprises of, regional cluster groups illuminate the common thread of how modern materials creation feeds upon the agglomeration synergies that are the fruits of tightly knit, multienvironment, high-technology communities.

The Southwest Region: Creation of New Clusters—The Dominant Company-Based Cluster Model [4]

In certain cases, the large corporation plays a key role in cluster formation. But if these companies help to kick-start the clusters into being, over time the cluster community takes on a life of its own and becomes a congregation of smaller firms, attracted or otherwise pulled into the environment created by the initiating firm and, over time, interacting with political (state and local), financial, and academic players. Such is the case with clusters that develop within the Southwest region. Within Texas and Oklahoma, in particular, advanced-materials clusters began to crystallize around such major corporations as Sematech, Halliburton, and Conoco-Phillips. These multinationals take on the role of initial benefactors and supporters of the cluster group. They do so for a variety of reasons: to secure an equity position in promising companies, upgrade the economic profile of their region, and secure a supply of advanced materials and technology for their own production processes. Within these regions, we find that advanced-material clusters can extend geographically over a wide area, even encompassing and extending out of the state. The following describes the evolution of three such advanced- materials cluster formations.

One of the fastest growing nanotechnology centers today is in the Dallas–Austin–San Antonio–Arlington area. Over the last decade, the Austin–San Antonio corridor has grown considerably as a high-technology center. By 2005, over 170 biotechnology and nanotechnology firms dotted the I-35 corridor from Temple to San Antonio, with the majority of these companies quite young (less than 15 years old) and small (less than 100 employees).

This cluster formed around the electronics research and development company Sematech, the leading research consortium for the global semiconductor chip industry. Within this region, Sematech nurtured an advanced-materials research center and a linked chip research and production facility ("national semiconductor foundry") to foster commercialization in nano- and biotechnology as well as chip manufacturing. Government (state and local) and academia eventually began playing important roles as they penetrated into and extended the cluster structure. Such was the case when the University of Texas began working with the Department of Defense (DOD) and Sematech in a large-scale strategic partnership to advance research in nanotechnology (the SPRING project). In short order, this high-tech group quickly attracted a number of smaller start-ups active in the field of advanced materials and the allied industries.

An even more extensive advanced-material grouping took root in and around the Houston area. The dominant firm operating in this area , the large Houston-based engineering company Kellogg Brown & Root (KBR, a unit of Halliburton) provided strategic support to such leading start-up companies as Carbon Nanotechnologies, Inc. (CNI) by providing needed capital, financing, equipment, and expertise. Founded in 2000, CNI licensed important patents from nearby Rice University in the manufacture of the advanced fullerenes and the so-called buckeyballs. As previously noted, these materials are single-walled carbon nanotubes that are very strong and can transfer electricity, heat, and other forms of energy for use in sensors, electronics, computing, lightweight materials, and drug delivery. CNI at first depended on local venture capital. Then, in 2001, CNI leased 6000 sq ft in KBR's technology center in Houston to commercialize its central nanotube process. At this time, it also signed an engineering services agreement with KBR, by which the latter provided conceptual and detailed design, fabrication, construction, and operations in support of present and future CNI technology. More specifically, KBR, tapping its knowledge of and experience in petroleum catalytic cracking, assisted CNI in the design, scale-up, and construction of a larger manufacturing reactor to produce 200 grams per day of buckeytubes at lower unit costs. These agreements, in addition to continued use of local venture monies, bolstered CNI during its critical early period. By 2003, CNI became the flagship nanomaterials firm in the region and the center of a growing advanced-technology cluster

As CNI grew, and the high-technology cluster of which it was part expanded, the company took a leading role in coordinating and expanding the cluster community. CNI soon became a dominant player and center around which other firms and financial and political agents congregated. CNI and its cluster group ultimately expanded their influence state-wide as they spearheaded the establishing of the Texas Nanotechnology Initiative (TNI), a state-wide consortium of Texas-based universities, industry leaders, investors, and government officials focused on bringing top nanotech companies, researchers, and funding into Texas and accelerating state funding of Texas' advanced-technology industries. In 2005, CNI's president, Bob Gower, helped form and was a member of TNI's Executive Committee. TNI considered Gower's presence on the Committee as absolutely crucial to its future success. Through TNI, the various clusters within the state work together to coordinate political, economic, market, and financial initiatives, and, in essence, became a more unified and integrated entity. This integration of different cluster groups helps secure and funnel more state funding and other subsidies into these regions and the firms operation there for technological development and economic expansion.

Spanning Oklahoma and Texas, a smaller but evolving cluster swirls around another well-known corporate giant. It was Conoco-Philips (the third largest integrated energy company and headquartered in Houston, Texas) that looked to revitalize the economic prospects of the region. The opportunity to do so came with the founding of the advanced-materials firm Southwest Nanotechnologies in 2001, which soon became a major regional producer and supplier of high-quality, single-walled carbon nanotubes. This firm and its technology evolved from research conducted nearby by the University of Oklahoma. Conoco, believing that the economic future of the region depended on the galvanizing potential of advanced materials, secured an equity share in the company in exchange for its support in terms of financial and intellectual capital.

By 2004, this nascent cluster complex expanded geographically and Southwest Nanotechnologies continued to exert its influence in the region's industrial growth and development. The company helped to propel business in this area by forming relationships with regional suppliers, in particular Zyvex Supply (Richardson, Texas). Southwest's' nanotubes were then incorporated into Zyvex's Nanosolve additive products, especially for polymer composite applications. About the same time, the company entered into a strategic partnership with Applied Nanotech, Inc. (ANI, Austin, TX). By this partnership, Southwest Nanotechnologies authorized ANI to represent it in the selling and distribution of single-walled carbon on a nonexclusive basis to ANI's contacts with major manufacturers in Japan. Additional companies began entering the area, in large part wanting to benefit from economies of shared knowledge, information, and technology. By 2005, Southwest Nanotechnologies stood at the center of an ever-widening interstate cluster.

The Upper New York State Region: Creation of New Clusters—The Politically Centered Cluster Model [5]

Similar to the Southwest, upper New York State is actively engaged in creating from the ground up leading high-technology clusters centered on advanced materials and nanotechnology. What distinguishes this region from other types discussed is the importance of the political dimension to cluster formation. Other cluster models certainly involve the political element to varying degrees, and other factors play vital and closely interrelated roles in New York. But it is clear that political action has been the galvanizing and coordinating force for high-technology cluster development here. In the case of the Southwest region, we identified three distinct, though ultimately linked, industrial groupings. The situation in upper New York State involves an expanding cluster centered in the Albany area.

In contrast to Silicon Valley, no preexisting intellectual or technical resource cluster existed here. Industrial activity that could act as context and support for emerging technology is largely absent as well. But the political system in place in the 1990s reasoned that it would be beneficial for the region, and through ripple effects the state as a whole, to work on creating an entirely new high-technology industrial complex from the ground up. This grass-roots model of cluster creation most strongly reflects the belief by public officials in the economic development potential of advanced technology.

With a price tag of between \$1 billion and \$2 billion, the state government has been guiding the hand of industrial and technological growth in the region since the late 1990s. New York State's strategy is to organize its high-technology complex around "core" centers for advanced materials and nanotechnological development, using State University (SUNY) colleges and universities as the center of it all. This center, in turn, would attract companies large and small into the yet-to-be developed area, enticed by growing R&D capability, tax and other financial incentives, and the relatively inexpensive housing market and general low cost of living.

In response to this welter of activity within New York State, interstate competition, especially between New York and Texas, has intensified. Upstate New York did not control the technical resources that Austin built up in the 1980s and 1990s upon which to build an advanced-material and nanotechnology cluster complex. On the other hand, Albany does have certain strengths including access to New York City, relatively traffic-free highways in the upstate area, research facilities at area companies and schools, newly emerging nanotech programs at Rensselaer Polytechnic Institute and Cornell University, as well as research hubs at such important R&D companies as General Electric (GE) and IBM.

New York State can access a deep financial pool as needed to entice companies into the region. The Governor's office has far more administrative and financial power in New York than it does in Texas. Although New York, like Texas, has its budget crises to deal with, New York (unlike Texas) is not required to balance the state budget and so can offer more funding to companies like Sematech who would bring technology and jobs into the area and attract around it a group of other high-tech firms. New York's Governor's office aggressively recruits high-tech projects. It offers as incentives designated "Empire Zones," where companies operate free of state income and property taxes for 10 years, as IBM does at its manufacturing plant in East Fishkill. State grants also cover a large portion of a company's worker training budget.

These incentives and strategies clearly accelerated cluster creation in the

Albany area. Although Sematech has headquarters in Austin, the company is sending increasing numbers of research personnel to Albany. Another company, Tokyo Electron Ltd. (TEL), which manufactures equipment used to make microchips, followed Sematech into Albany. The company has sent hundreds of researchers to work in the $300 million R&D center at the State University of New York at Albany (SUNY), New York State won out over two other communities that competed for Tokyo Electron. Albany is TEL's first dedicated R&D center outside of Japan.

With Sematech and TEL as the beginning, New York State strategy hinged on developing a comprehensive nanomaterials and nanotechnology complex that feeds into the needs of these two flagship companies. This has indeed been happening. The company Nanocs International Inc., which makes metal semiconductor layers for computer chips, relocated out of New York City and moved to Albany. In doing so, it rejected opting for relocating to Texas (which had also vied for its business) to be near to the growing constellation of advanced-technology companies around SUNY.

To facilitate the clustering process, the State of New York innovated the administration of advanced technology complexes with the creation of NanoTech Resources Inc. (NTR). NTR provides the strategic headquarters of the State of New York's nanotech research and education programs. Essentially, NTR is the umbrella organization for Albany NanoTech, SUNY at Albany's School of Nanoscience and Nanoengineering. Whenever new firms and organizations come to the site, NTR helps to form partnerships with companies, seek funding sources, and act as liaison in establishing working relationships between business and academia.

The swift growth of this cluster has been duly noted by the business and high-technology industries. In 2003, the Forbes/Wolfe Nanotech Report named the then Governor (Pataki) one of the nation's top ten leaders in technology development; and the important trade journal *Small Times Magazine* of March/April of that year ranked New York seventh as a "small-tech hot spot."

Since 2003, Albany's cluster has continued its forward expansion. It also engrosses a growing number of disciplines, including biotechnology. The cluster includes the Center for Excellence at SUNY Albany, as well as centers for Bioinformatics in Buffalo, Photonics in Rochester, Environmental Systems in Syracuse, Biotechnology in Troy, and Information Technology on Long Island. New York State supports the integration and coordination of these various centers through policies, directives, and funding support. It also acts to entice subsidiary firms—financial, marketing, economic, and suppliers—into the region through strategic incentives. In 2005, Honeywell, attracted to the regions by these incentives and by the prospect of partaking in the growing economies offered, announced it would invest $5 million over five years in equipment and research in Albany NanoTech. Honeywell's plans include locating laboratories and researchers at the center to work on next-generation materials for the semiconductor industry. For instance, the company is developing metal precursors and other materials related to atomic-layer deposition (ALD), a semiconductor manufacturing technique that deposits a single layer on a chip that is only one atom or one molecule thick. This is an important technology, especially as chips become increasingly smaller.

In addition to large companies like Honeywell, the main goal by 2012 is to entice the small, more innovative firms into the Albany-based cluster. It is clear to observers at the scene that a cluster effect has started to build momentum in the region with spin-off companies attracted to the growing synergies [6].

Evident Technologies is a case in point that allows us to see how such start-ups enter and, in a systematic, iterative process, integrate into, a multidimensional cluster environment. Evident Technologies, formed by employees from Lockheed Labs, began operations in 2000, specializing in bio-nanotechnology. In particular, It found methods to produce semiconductor nanocrystals (quantum dots) in commercial quantities for biosensors, optical transistors, switches, optical computing, LEDs, and lasers.

The real growth of the company came only after its move into the Albany region, induced by the offer of New York's financial incentives administered jointly through the Empire State Development Board of Directors and the Small Business Technology Investment Fund. In exchange, the state received an equity interest in the company. Beside funding, advantages for a start-up company like Evident in this arrangement included access to a network of contacts with other companies in the region; advice on strategy, personnel, and hiring; and boasting points with other investors on business plans and prospectuses in being able to claim that the State of New York invested in Evident and backed its development plans and projections. Within a short time, the company began securing funding from local investors. By 2003, the company had closed on nearly $4 million in funding from various local sources, public and private.

As Evident integrated into the region, it took advantage of the economies that exist there with various actors operating in the cluster. Beside funding sources, Evident established close ties with both SUNY Albany and then Cornell University (Ithaca), thus benefitting from knowledge transfer and licensing agreements. It also participated in state-funded incubator facilities, notably the INVEST incubator system created by state funds (nearly $3 million) in close cooperation with nearby Russell Sage College.

Evident also formed development and market agreements with companies operating within the cluster group. These agreements catalyzed important resource transfers of technology, knowledge, and skills. Evident formed such an alliance with nearby IBM. In this arrangement, Evident provides advanced materials in exchange for IBM furnishing management and technical consulting services. In 2004, Evident entered into agreement with the company Upstate, a leader in cell signaling products for life science research and drug discovery. By this arrangement, Upstate incorporates Evident's quantum dots (EviTags) into its own products and systems. The success of Evident in entering into and thriving within upper New York State has been a positive incentive for other companies to join the high-technology cluster.

California: The Revitalization of Maturing High-Technology Cluster Model—The Central Role of the University [7]

California exhibits a unique model of cluster growth. Here, newer advanced-material cluster groups form around the older layer of Silicon Valley's clusters of the 1970s to 1990s. This means that business developers and investors hope to extend Silicon Valley's expertise in semiconductors and electronics to the new technologies. However, problems loom for the Valley. Not least of these are the usual difficulties that come with success: overpopulation and the high costs of living, which threaten to choke off further growth as companies decide to relocate to less developed areas.*

Despite these warning signs, Silicon Valley continues to show strong signs of growth into the first decade of the twenty-first century. The universities and government research laboratory continue to fuel so much new technology in the region. The work of these universities and government labs supports advanced-materials sciences and engineering. For example, an important company, Molecular Nanosystems, was founded in 2001 by an Associate Professor of Chemistry at Stanford to develop and produce carbon nanotubes (both single and double walled) in large arrays and networks using a scalable chemical vapor deposition technique. Molecular Nanosystems rooted itself in the area in order to tap into the technical, financial, and market synergies that exist. In 2002, the company set up a joint venture enterprise (with General Electric), called GE GRMN, to develop new materials for advanced field-emission devices being developed within Silicon Valley.

In a similar manner, Nanosys, founded in 2001 and based in Palo Alto, secured important rights to intellectual property developed by scientists and engineers at UCLA and Berkeley, as well as the Lawrence Berkeley National Labs. From these beginnings, Nanosys developed advanced systems composed of such materials as nanowires, nanotubes, and nanodots, including sensors, silicon solar panels, memory chips, and thin films for flexible electronic displays.† As with Molecular Nanosystems, Nanosys integrated closely into the Silicon Valley complex, making creative use of other technology companies (such as Hewlet Packard), law firms, venture capitalists, and advanced-material enterprises. Nanosys also leveraged its growing presence in Silicon Valley to extend its influence over the state-wide technological community such as organizing the Northern California Nanotechnology Initiative (NCTI). Through political influence at the sate (and federal) levels and the extensive reach of its members through Silicon Valley's industrial, academic, and

*The Silicon Valley region has been cutting education spending and raising workers' compensation costs (and therefore labor costs). Also, the high costs of housing and transportation could, over time, turn potential entrants away.

†The importance of the company is indicated by the fact that Nanosys was named 2004 Technology Pioneer by the World Economic Forum. Nanosys won *SmallTimes Magazine* Best of Small Tech Top Researcher award and was invited to attend signing ceremony for the 21st Century Nanotech Research Act by President Bush.

financial communities, NCTI, led by Nanosys and other companies, helps to coordinate the activities of the different advanced materials firms and organizations by linking together complementary advanced-technology companies and helping to find venture capital for promising start-up firms operating, or looking to set up shop, in the Valley.

Mid-West: The Revitalization of Old Industry Model [8]

The extension of an existing cluster like that of Silicon Valley is not an option for the Midwest. As was true of upper New York State, the Midwest region did not develop extensive high-technology clusters in the 1970s and 1980s. The region suffered from decaying "smokestack" industries such as the chemicals and plastic processing. But, in contrast to the Albany area of upper New York State, which did not previously (i.e., prior to the 1990s) foray deep into the materials industries, the industrial history of the Midwest engrossed active and once-dominant chemicals and metals sectors. In the 1990s, research conducted in the Midwest began to focus on ways in which to use the region's knowledge in these older materials to help spur the growth of nanomaterial clusters. Knowledge of these older industries, often embedded in university engineering departments, readily transferred into work on the most advanced materials.

By 2003, such cluster formation appeared on the industrial radar. These clusters incorporate a growing number of diverse actors, firms, and agencies, and more closely integrate the important players. According to *Small Times,* over half the companies in the United States involved in nanotechnology are now located in the so-called "nanobelt," which runs from Illinois to Massachusetts. Companies such as Nanofilm and Nanophase Technologies, with strategies to exploit and update such older industries as chemicals and plastics, help to establish advanced-material-technology industrial groups in Ohio and Illinois, respectively.

Headquartered in Cleveland, Ohio, Nanofilm develops and commercializes ultrathin films possessing good durability and clarity and used, among other things, as coatings for eyewear lenses and for displays in cell phones, touch screens, and cameras. The company is one of the oldest nanomaterials firms in the United States, and evolved out of work begun at Case Western Reserve University. This research agenda has helped revitalize Ohio's large but stagnant polymer industry, which had not introduced an important innovation in recent years.*

This research aimed to modify and improve traditional plastics through advanced thin-film coatings. The company initially depended on local funding. This included a network of contacts and local venture capital, such as Ohio's Maple Fund. Nanofilm's first important customer was the Cincinnati-based LensCrafters, followed by automotive and transportation firms in the Ohio area.

Since the 1990s, these beginnings began to spawn a growing network of indus-

*Experts in the industry said no new plastics were going to be commercialized as it would be too expensive to introduce and, therefore, much easier to formulate existing materials than to introduce radically new materials.

trial, financial, and distribution companies and firms. The leading companies within this regional network, including Nanofilm, worked to induce Ohio to integrate its high-technology sectors around advanced materials. Specifically, Nano-Network was formed by scientists, entrepreneurs, and financiers to improve and expand nanomaterial technology research and commercialization activities and capacities in the Northeast Ohio area. This effort resulted in greater cohesion within Ohio's high-technology regional cluster. These clusters have taken root in Ohio because the region's long-standing expertise in ceramics, metals, and polymers all congregated here with ample opportunities for cross-fertilization. In addition to Case Western Reserve, which houses the Center for Micro and Nano Processing (CMNP), the other players in Ohio's advanced-materials industry include Kent State University, Ohio State University, as well as Wright Patterson Air Force Base and Glenn Research Center.

Since the late 1990s, this regional network has been growing and becoming more integrated and interdependent. Personnel transfer from one company, agency, and organization to another is robust and causes a vigorous flow of information, skills, and even technology from one place to the next. This activity and mobility generates a great creative surge that is enhanced by the fact that the region traditionally is rooted in material technology and know-how. The technical and economic impact of this cluster is such that, according to *Small Times,* Ohio in 2006 already had the tenth-best advanced-materials industry nationwide.

EUROPEAN CLUSTER FORMATION

As the different models show, the United States creates a wide range of cluster types that lead to creative innovation. Gone are the days when natural resources supply the galvanizing force for cluster formation. A variety of triggering mechanisms and organizing forces come into play within the United States. Europe presents a very different picture.

The European Scene

The concept of cluster synergies is certainly not the preserve of the United States. Europe also has been attempting to increase its competitiveness in the world through cluster formation. Industrial clusters exist in Europe, and are even growing in number. Indeed, some of these are rather extensive and thriving. Nevertheless, they tend to be organized around the more traditional or older industries, such as textiles, machine tools, metal forging, sports goods, furniture, and soap, and so do not lend themselves to significant new economic growth or national competitive advantage.

High-technology clusters of the depth and breadth found in the United States are fairly rare in Europe. Moreover, they are not particularly like typical American advanced-material cluster groups. They tend to be top-heavy with a bias that favors control and funding by big government and corporations, and often are oriented toward large-scale science projects coming out of theory-oriented universities. These clusters tend to specialize in one field or another, thus defeating the creative impact of the cross-fertilization of ideas, technology, and personnel between disciplines within a single, integrated cluster group.

The case of France illustrates this point [9]. The emphasis in France is on guidance and funding from the European Union and the national government. In 2005, the French Prime Minister announced funding for six industrial clusters and 61 "competitiveness" clusters. Funds for this well-meaning program doubled from 750 million to 1.5 billion Euros. Along with this funding, the federal and national governments attempted to engineer the process from the top by organizing clusters according to scientific and technical specialty. These clusters include the Rhone–Alpine Regional cluster, focusing on health care; the Bordeaux-Toulouse cluster, emphasizing aerospace; the Paris cluster centered on transport and navigation; the Southern France cluster, specializing on telecommunications; and the Grenoble Regional cluster, targeting nanotechnology.

In engineering the growth of these clusters, the French government places heavy reliance on its university and research network, consisting of four major universities, nine public research institutes, and nearly two hundred laboratories. For example, the University of Montpellier advanced-materials laboratory spawned the company Nanoledge, Europe's first large-scale producer of carbon nanotubes. The strategy was to position Nanoledge as a central player in one of France's high-technology clusters. In truth, the start-up depended almost exclusively on national grants and theoretical research through a "narrow collaboration" with Montpellier, a well-known science-based research center.

Cluster formation in Switzerland and Germany demonstrates a similar pattern [10]. These countries also place heavy reliance on federal and national funding and leveraging their scientific excellence into new-materials technology. Nationally, Switzerland allocates 0.8% of its GDP to basic research, almost two times that of the United States and Japan. There is also much closer collaboration between government and academic institutions than in the United States. Basel has become one of Switzerland's most important high-technology clusters. Such a model of cluster creation applies in Germany, where the Federal Ministry of Education, Science, Research, and Technology (BMBF) provide substantial national support for advanced materials and nanotechnology programs. New cluster groups tend to emerge from German scientific institutes. One of Germany's largest projects is CESAR, a $50 million science center in Bonn, closely linked to Germany's most prestigious research organizations. The Fraunhofer Institutes, Max Plank Institutes, and several universities have also formed "centers of excellence" in the advanced-materials field.

Moving northward to the Scandinavian countries, Sweden's industrial clusters form closely around the university science departments such as in Lund, Goteborg, Stockholm, Uppsala, Linkoping, and Umel. An especially active region is the Øresund area, containing the cities of Lund, Malmo, and Copenhagen, twelve universities, and a growing number of research-intensive companies.

In its attempts at creating competitive clusters, the Dutch also endeavor to position their science and university research centers as the engine of industrial expansion. Holland's recently created NanoNed network is a consortium of scientific-research-based universities and institutes. NanoNed consists of national investments in experimental facilities, scientific research, and dissemination of scientific knowledge, which, as the following passage indicates, the Dutch consider the three "main thrusts" of its nanotech initiative.

> This . . . approach ensures the use of newly developed knowledge across the complete scientific field, and stimulates swift implementation of innovations across fields of physics, chemistry, and the life sciences. . . . [T]his approach enables the consortium to ensure that all essential scientific subjects, necessary to create the new generation of nanotechnology-based products, are present and available . . . for all partners. . . . [T]he NanoNed Programme focuses on its science and investments in equipment and knowledge dissemination. [11]

So we see that, in stark comparison to the play of agglomeration economies in large, diverse, engineering-based development-from-the-bottom United States cluster structures, those in Europe depend too greatly on central funding, a foundation in "big" scientific ideas, and specialization of function. This prevents the fashioning of that integrated seamless web so critical in linking together and putting into play productivity-enhancing technology.

The following case study examines the evolution of the Oxford technology area in the United Kingdom. It offers an excellent counter-example to the U.S. situation. As Europe's generally accepted premier high-technology cluster, it presents to us the upper limit of what such a European complex is capable of achieving in the first decade of the twenty-first century and can be then compared to the impressive contributions that U.S. high-technology clusters make to economic growth.

Case Study: Oxford's High-Technology and Advanced-Material Cluster [12]

Oxford and Cambridge are generally considered the two most consequential technology clusters in United Kingdom and, indeed, in Europe as a whole. For example, they contain the majority of the United Kingdom's biotechnology companies composed of renowned research universities and geographical concentrations of companies, including relatively new start-ups (often spin-offs from the universities) living side by side with larger, more mature companies. Proximity to established pharmaceutical and biotechnology companies provides the newer SMEs with partnering opportunities for product development and manufacturing, marketing, and management expertise.

Oxfordshire, which is strategically located in the southeastern part of the country, is the most prosperous region in the United Kingdom. Its economy is influenced by the city of Oxford, with its significant representation in important sectors concerned with education, healthcare, the motor industry, publishing, R&D, and tourism. Known as the U.K. diamond region, the complex stretches one hundred kilometers in a radius around Oxford. This represents one of the largest agglomeration of scientific research activity and talent in the world. Its range of research includes nanomaterials, genomics, biometrics, Internet services, wireless, and biotechnology. The region is ranked twenty-second out of 77 European regions in terms of its GDP output. Europe considers Oxfordshire as one of its leading centers of enterprise, innovation, and knowledge. At the end of 2001, the region contained over 1400 technology-based firms dealing directly or indirectly with materials, employing some 37,000 people. Between 1991 and 2000, Oxfordshire experienced a faster rate of growth in high-tech employment—about 20%—than any of the other 45 English counties.

The situation within the Oxford region is actually less impressive than the above statistics show. If we look at *all* manufacturing employment in the Oxford region in the 1980s and 1990s we find that, between 1981 and 1998, there was an overall loss of 11,200 jobs. This loss was, in large part, a result of closure of the Cowley Car Works and the impact this closure had on associated industries in Oxford. In other words, the high-tech industries that formed within the area did not create sufficient jobs to offset the loss of a plant closure in a traditional manufacturing sector. We may compare this situation with what we know of US high-technology cluster dynamics that we examined, whereby technology creation through SMEs, supported by a thriving, dynamic, and specialized venture capital community, replace or revitalize industrial regions, resulting in robust employment growth and expanded economic activity throughout the region. In comparison, during the decade of the 1990s, it is estimated that total employment within the most successful U.S. high-technology clusters grew in the range of 60%–120%.

The evidence shows clearly that U.K. officials and business groups themselves understand, or at least are aware of, the shortcomings of its most important cluster group and its economic performance, especially in comparison with those of the United States. The Oxford City Council noted in 2002 that its premier "Oxford Business Park" has fallen far short of replacing the number of jobs lost. As Oxford government officials admit, there has "not . . . been a net increase in *total* employment, even viewed in the short term, as a result of the Development" [13]. The Council has expressed the great need to create larger and far more coherent clusters that closely integrate related and complementary companies and that establish closer ". . . links between business and the science and research base in order to encourage U.K. business to compete successfully in world markets" [14]. The failure of the Oxford region to meet its economic potential is also reflected in the fact that the South East England Development Agency (SEEDA), as late as 2004, still sought (i.e., had not yet achieved) promotion of sustainable economic development in the region and the strengthening and growth of established and emerging businesses. Similarly, the Oxford Economic Partnership continues to look for ways to make the region globally competitive in terms of employment and productivity.

Recent studies conducted by U.K. organizations note that problems abound throughout the country's advanced-material and high-technology cluster groups that limit the realization of full economic potential. These regions include London, Central Scotland, Yorkshire, and parts of Wales. According to the 1999 U.K.

biotechnology cluster report, "The UK leads Europe in advanced materials and biotechnology [cluster achievement], although it is still some way behind the United States." The report goes on to list a number of barriers and disincentives that remain to the ". . . effective exploitation of the U.K. [knowledge] base compared to the United States. . . . [Compared to the United States, high-tech] cluster networks in the United Kingdom are still in their infancy, . . . [accordingly] the United Kingdom can and should learn from the experience of American clusters" [15]. And, indeed, United Kingdom groups have begun to form partnerships with American agencies and organizations to do just that. For example, the United Kingdom's Eastern Region Biotechnology Initiative established a partnership link to the Massachusetts Biotechnology Council. Such international partnerships are likely to grow as Europe attempts to transfer and adapt American methods of cluster development for its own advantage.

Fragmentation and Critical Mass: Clusters and The Multienvironment

Ultimately, the issue here is one of fragmentation. Within the United States, a wide variety of high-technology clusters actively pursue advanced-materials (and advanced-technology) development and commercialization. Although different types of U.S. advanced-material clusters exist, they have common traits. In the first place, large government, whether national or supranational, plays only a secondary role in terms of funding, planning, and guidance. Rather, local governments and often locally based sources of angel and venture capital actively invest in promising firms and technologies. The federal government, held back by red tape and conflicting interests in different parts of the country, is not particularly effective in selecting the best firms, nor does it have an impressive track record in guiding commercial technology. Most importantly, since new-materials technology arises within a regional cluster context, it is vital that that region and that locality, and their economic, financial, and technical profiles, be clearly understood. Thus, state and local governments—those entities "closer to the ground" and with a greater stake in job growth and economic development within that area—are the ones more appropriate for successful cluster formation, as we saw in the case of New York State.

Further, U.S. advanced technology clusters do not ally themselves too closely with the science departments of universities and research institutes. Practicality and utility are the watchwords in American technical and economic progress. Flexibility, on the one hand, and interlinkages, on the other, are the vital components of American cluster groupings. As we see, American clusters engross a number of diverse actors that nonetheless interrelate closely with one another in a synergistic fashion. If in some cluster groups certain entities—whether large established corporations or state and local governments—play a central coordinating role in cluster formation; they cannot go it alone. Rather, they serve as motivators and catalysts in bringing together and linking financial, technical, scientific, market, and political forces within a region. This is certainly crucial because innovation depends less on pure science and research than on the constant overlap and dialogue between these interdisciplinary forces. Clusters succeed then only if all these components actively communicate with one another simultaneously, creatively, and on a day-to-day basis.

The basic theme that suffuses America's successful high-technology growth, then, dances to the rhythm of coherence and integration. In industrial clustering of high technology, a proper specialization and division of labor is necessary to achieve critical mass or size, allowing employment of specialists of various types to be linked together in an integrated structure that is necessary for further economies and growth. Successful commercial clustering requires the simultaneous and coordinated actions of a number of factors, including political leadership, R&D centers and incubators, capital availability (area banks), specialized services, networking, and, most importantly, entrepreneurial energy and innovation. In general, U.S. clusters show a greater dynamism in capturing, marshalling, coordinating, and harmonizing these often conflicting forces than other countries and regions, and that has led to greater practical applications and economic impact. This impact touches the local economy to be sure, but also, through the ripple effect of indirect and induced benefits, percolates throughout the national economic fabric.

In the European Union, a different pattern emerges. Here, fragmentation rules and seriously blunts the force of agglomeration economies that the cluster environment potentially offers. The European Union lags behind the United States in developing commercial technology from cluster efforts. European clusters tend to be too incoherent, consisting of separately operating unconnected small- and medium-sized enterprises, and geographically dispersed. Thus no one region has achieved the critical mass necessary to meet what has been termed the "international visibility" required.

This persistent problem of cluster fragmentation is in part due to regulatory and fiscal policies of the European Union and its national governments. These governmental entities, being more overarching and strategically oriented, do not effectively design cluster plans at the local level, where, after all, successful agglomerations must take root in order to grow. The case of New York State shows the impact state and local authorities, which are "closer to the action," can have on cluster formation for critical localities. But, more fundamentally, Europe's problems point to historical issues, such as the national nature of research resulting in little cooperation among member states. Indeed, European companies are most likely to enter into collaborative R&D agreements with U.S. firms than with other European companies. It is also well to remember that the interface between the E.U.'s public research system and European industry is not sufficiently developed, which also leads to fragmentation. This means a lack of open and transparent dialogue between all stakeholders concerned and between countries and regions. Without this dialogue, the E.U.'s efforts at coordinating advanced material technologies within the cluster environment remain confusing, and often lead to conflicting views between the European Union, its member countries, and outside trading partners, such as occurred in the case of genetically modified foods.

Another and related problem is the loose coordination between companies in a cluster. In this sense, European clusters appear less cohesive and well defined than in the United States. European clusters consist of a number of larger, mature, and traditional firms that typically restrict information exchange with other neighboring companies. This deep-seated tradition of secrecy means that European clusters tend to be "contract" rather than cooperation driven. They do not accrue interactive benefits to the same degree as U.S. industrial complexes and, therefore, cannot reach the necessary dimensions to be recognized as promising clusters by the international community. European clusters tend toward specializing in one industry or another, rather than (as in the United States) embracing a congregation of different but intertwined industries. Within the European Union then, we find an inordinately weak convergence of materials, electronics, and telecommunications. The latter is concerned mostly with basic network infrastructure and service delivery, and operates separately from new-materials development. This can only mean that critical supplier relationships do not seem to have survived or emerged as part of the cluster structure.

The issue of fragmentation arises also between knowledge and application, and theory and commerce. Even in those situations in which Europe attempts high-technology cluster formation, problems persist. A difficulty in creating such clusters in Europe is the deeply rooted belief that technology arises out of pure scientific research. Not surprisingly, then, Europe's clusters tend to form around science-oriented universities. For instance, a core of science–technology enterprise has been gathering around Oxford University. Here, the European tradition that sees profits as a taint on science means that the university views clusters as a way of creating knowledge efficiently rather than as a commercial venture. In contrast to U.S. cluster performance, the U.K.'s diamond region remains scientifically oriented with commercial possibilities, and the job creation that would go with it is still untapped. Here, as in other European regions, there is great pressure to publish research results, the sine qua non of scientific status. But this is anathema in real-world industry, where keeping one's council in new discovery secures competitive advantage. In Europe, unlike the United States, status-linked rewards for commercial success do not match those garnered in scientific achievement. In keeping with this sort of prioritization, important transition points for science–commercial interchange, such as technology transfer offices within universities, have not yet reached the same level of activity or sophistication in Europe as is currently evident in the United States.

Europeans do appear to be gaining a greater insight how to encourage entrepreneurship and coordination between academia and industry. A common theme in many recent E.U. reports on building successful high-technology clusters is that young researchers in Europe often lack opportunities to build the skills needed for commercializing research. They stress repeatedly the need to train undergraduates and graduate researchers in management and entrepreneurship. This recent realization comes in the wake of an entrepreneurial "brain drain" from Europe to the United States. As one British study concludes, a growing number of "British entrepreneurs [have been attracted] to work in the U.S. [advanced technology] industry, attracted by American (can-do) spirit and liberal exercise of share options and tax advantages." Europeans feel that their approach to advanced technology is not working and look to programs from schools such as MIT as their model for how to structure R&D so as to maximize chances for commercialization and market placement within expanding clusters. Arguably, the most important discontinuity within European clusters is between companies and their sources of funding, especially venture capital. Within the United Kingdom, for example, European analysts conclude that companies and investors are not located sufficiently close to each other in clusters, resulting in a widening gap in the amount of equity financing available for nanotechnology and advanced-materials companies. This is in stark contrast to the close relationship existing between start-up companies and venture capital within the United States [16].

These problems within European clusters reveal themselves in a very palpable manner when we compare the European and U.S. situation. The U.K.-based consultancy firm Robert Huggins Associates compiled a list of cities that are the most important knowledge-based and technology clustered regions. Ranked against 19 knowledge-economy benchmarks, including patent registrations, R&D investment, education spending, and access to private equity, the report shows that E.U. regional clusters continue to struggle to bridge the knowledge gap that would enable them to reach critical mass and compete with the United States globally (see Table 9.1). Accordingly, important industrial clusters in Europe were still concentrated in just a few regions and are not developing as fast as those in the United States. Their rankings show the domination of U.S. regions clusters within the world's high-technology-cluster communities.

As a result of these various forms of fragmentation, Europe's nano- and ad-

Table 9.1. World (advanced-material) knowledge competitiveness index ranking, partial list, 2004

City	Country	Index	Ranking
San Francisco, CA	United States	259.0	1
Boston, MA	United States	230.4	2
Grand Rapids, MI	United States	197.3	3
Seattle, WA	United States	196.3	4
Hartford, CT	United States	195.4	5
San Diego, CA	United States	192.5	6
Rochester, NY	United States	191.8	7
Sacramento, CA	United States	183.0	8
Austin, US	United States	183.9	9
Minneapolis, MN	United States	180.5	10
Stockholm	Sweden	170.7	15
Helsinki	Finland	154.7	19
Ile de France (Paris)	France	133.5	34
Tokyo	Japan	123.8	38
Shiga	Japan	123.1	39
South East	United Kingdom	119.8	40

Source: [17].

vanced-material clusters are not as dynamic as the leading U.S. clusters in New York, New England, Texas, and California. Comparisons of commercially relevant output prove instructive as well, and appear to confirm the general conclusions of Robert Huggins. A study conducted by the EU Commission found a significant gap in performance in technology transfer enterprises, or TTIs, involving nanotechnology and advanced materials. TTIs operate in clusters and are entities that generate "science and technology" (S&T) knowledge. Such knowledge, when transferred into the commercial world, creates new products and processes that spur productivity and economic growth. Patents, of course, are one measure of the real-world usefulness of TTIs. The more effectual the cluster within which a TTI operates—meaning more tightly integrated and coherent to engender vital economies that facilitate technology transfer—the greater the useful output of the TTI in particular and the cluster, made up of a number of TTIs, as a whole.

In fact, as Table 9.2 shows, the study found that the number of patents filed or issued by European TTIs is much lower than those of the United States. Half of the European-based TTIs filed no patent at all in 2002. The average number of active licensing contracts amounted to about 120 in the United States, while European TTIs average only 17 active contracts (and half of the European TTI's had only two or less active contracts). In 2002, the average license revenues per European TTI amounted to less than a quarter of the revenues of the United States (4.7 Euros versus 20 million Euros). Further indication that European TTIs associated with industrial clusters are not very outward oriented nor particularly successful in their marketing and communications strategies reveals itself by the fact that half of the TTIs had fewer than 21 clients compared with 80 to 100 clients typical for the United States. And the United States created more firm spin-offs per TTI or per university than did Europe. As one observer on the European high-technology scene notes, Europe severely underperforms in its transfer activities compared to the United States. "Even if one recognizes that the European and U.S. patent systems are not fully comparable, these figures suggest a distinctly lower level of activities of European TTIs" [18].

Given these differences, it is not surprising that In Europe, when industrial clusters flourish as commercial entities, they do so under the leadership of U.S. companies. For example, a new R&D facility in Grenoble that currently employs 450 en-

Table 9.2. Comparison of cluster-based technology transfer performance indicators in the United States and the European Union (per TTI), 2002

Output	United States, mean	Europe, mean
Number of patents filed	35.8	6.2
Number of patents issued	16.8	5.8
Number of active licensing contracts	120.2	17.1
Revenue from licenses (thousands of Euros)	10,173	507
Number of spinoffs	2.1	1.6

Source: [19].

gineers is expected to create 5000 jobs throughout a number of industrial sectors in the region around the city. As such, this project represents France's largest industrial investment. At center of this complex is a joint partnership between STMicroelectronics, Motorola, and Phillips.

The European Commission has taken the measure of the differences between the United States and Europe in industrial clustering and concludes that in Europe:

> . . . there is a potential conflict between academic achievement criteria and commercialization activities. . . . Only a minority [of researchers] see commercialization as important. . . . Low visibility to industry reflects insufficient outward-orientation and failed communication strategies. . . . The small average number of clients per TTI reflects weaknesses in marketing. . . . [Future success for Europe in high-technology clustering] requires proper understanding of the market and an adequate pricing and communications strategy. . . . [Most of all it] requires a bundle of technical, legal, and business administration skills [and] an ability to establish and maintain networks via communication. [20]

The ability of a country to maintain integrated, fast-growing clusters allowing free flow of information, personnel, and technology between members in the complex depends on the existence, creativity, and energy of broad-viewed individuals who creatively combine technical, legal, business, and regulatory forces. These wide-ranging actors, known as gatekeepers, span and serve as conduits for the interchange between the different disciplines. They create porous boundaries between science, technology, commerce, and government. Although they are not new on America's technology scene, they never played a more important role in society than they do at the dawn of the 21st century. In preventing segmentation and fragmenting of industrial cluster groups, they are the very arbiters of twenty-first century economic growth. The following chapter discusses these individuals and their roles in advanced-materials technology and national competitive advantage.

REFERENCES

1. For discussion of clusters see Porter, M. (1998), "Clusters and the New Economics of Competition," *Harvard Business Review,* Vol. 76, No. 6.
2. This idea of the "regionality" of innovation within clusters is a common theme in many of the reports and studies mentioned in this chapter. Europeans appear to believe that their slower progress in new advanced-materials technology is due in large measure to their too great reliance on orders, direction, and funding from central governments that do not have the same specific and intense incentives to innovate as do local entities. The latter, E.U. studies show, instill rough-and-ready entrepreneurship and risk taking in tackling impinging, real-world barriers, which leads to truly new solutions.
3. This interrelatedness comes through very clearly in a study of the evolution of regions such as the Niagara area in the late nineteenth century, and Silicon Valley in the twentieth century. See for example Trescott, M. (1981), *The Rise of the American Electrochemicals Industry, 1880–1910,* Westport, Connecticut, Greenwood Press, pp. 5–15 for a

study of the Niagara region, and Adams, S. B. (2005), "Stanford and Silicon Valley: Lessons on Becoming a High-Tech Region," *California Management Review,* Vol. 48, Issue 1, pp. 29–51.

4. *Houston Business Journal* (2001), "Kellogg Brown & Root, Carbon Nanotechnologies to Develop Buckytube Technology," September 27; Ladendorf, K. (2003), "With Bid to Keep Sematech, Perry Wants to Make Texas a Nnoscience Hub," *Austin American-Statesman,* May 1.; Nanotechnology Foundation of Texas website (www.nanotechfoundation.org) (2004); Stuart, C. (2002), "Texas: Its Small Tech Machine Well-Oiled," *Small Times;* Texas Nanotechnology Initiative Website (www. texasnano.org) (2006); NanoTechCafe (www10.nanotechcafe.com) (2004), "Carbon Nanotechnologies, Inc. Supports Governor Perry's Proposal for Texas Emerging Technology Fund," December 16.; ForRelease.com (2005), "Texas Nanotechnology Initiative (TNI) Elects Dr. Zvi Yaniv (CEO, Applied Nanotech Inc. in Austin, TX) to its Executive Committee and Board of Directors to Help Drive Commercialization of Nanotechnology within Texas," July 21; Winstead Sechrest & Minick (2005), "TNI Forms Executive Committee to Focus on Government Lobbying and Engaging Large Texas Corporations for Texas Nanotechnology Industry," Press Release, Austin, Texas, May 4.; *Austin Business Journal* (2005), "Group to Lobby for Nanotech Funds for Austin, Texas (May 4); Gardner, E. (2003), "Youthful Nanotech Company Grows Up—And Starts to Play with the Big Kids," *Small Times,* October 20; Houston Chronicle (2005), "CNI Points to Patents, TVs for Growth," March 8; Stuart, C. (2002), "CNI Has the Brains, the Cash, Now All It Needs is the Market," *Small Times* (www.smalltimes.com). July 29.
5. D'Errico, R. A. (2003). "Albany Nanotech Deal Maker Tells His Story," *Business Review,* Vol. 30, No. 31 (November 7); D'Errico, R. A. (2003), "German Microchip Maker Mulls Sending Researchers to UAlbany," *The Business Review,* April 7; D'Errico, R. A. (2001), "Two Area Businesses Approved to Get Tech Funding, Evitech Only One to Get it in Albany–Upsate Area," *The Business Review,* December 21, Albany Nanotech Website (2006), News (www.albanynanotech.org/News.cfm); D'Errico, R. A. (2002), "NanoBusiness Alliance to Open Albany Hub," *The Business Review,* September 20;Mokhoff, N. (2003), "Semiconductors: New York Planners Chart Course for 'Technopolis'," *EE Times Online,* April 30, 2003; State of New York, Executive Chamber (2002), "Nanotech Firm to Invest [and] create New Jobs . . . ," Press Release (Albany, New York), September 13; Rensselaer Polytechnic Institute (RPI), Lighting Research Center (2004), "LRC to Develop White-Light Illunimation System Using Quantum Dot Nanomaterials with Evident," Press Release (Troy, New York), July 15; State of New York. Office of Science, Technology, and Academic Research (2005), "Albany Center of Excellence Given Top Ratings by Prominent Nanotech Magazine," Press Release, Albany, New York, June 10; Smriti, J. (2003), "Michael Connolly —Big Things Expected from Nanotech Firm," *Rochester Business Journal,* Vol. 19, No. 16 (July 25), p. 10.; D'Errico, R. A. and Phillips, M. (2003), "Lucent Scouts UAlbany, RPI for Nanotech Work," *The Business Review,* July 7; SmartStart Venture Forum (2002), "Company Profiles: Evident Technologies, Inc." (Albany, NY); Evident Technologies, Inc. (2003), "Evident Technologies: Recruiting and Financing Technology Businesses," PowerPoint Presentation (Troy, New York); Evident Technologies, Inc. (2002–2006), Press Releases (Troy, New York).
6. See for example Sickinger, T. (2003), "Luring All Things Nanotech," *The Oregonian,* July 13.
7. Huang, G. T. (2003), "Nanosys and Quantum Dot," *Technology Review,* Vol. 106, No.

8 (October), pp. 98–109; *Knight-Ridder Tribune Business News* (2004), "Nanotechnology Start-ups Flourish in California's East Bay Area," February 1; Fitzgerald, M. (2003), "Silicon Valley Faces Eastward to Catch Government Gold Dust," *Small Times,* September 4; *Silicon Valley/San Jose Business Journal* (2003), "Silicon Valley May be Headed Down the Nano Tube," April 14; Adams, S. B. (2005), "Stanford and Silicon Valley: Lessons on Becoming a High-Tech Region," *California Management Review,* Vol. 48, No. 1, pp. 29–51; Forman, D. (2003), "Fully Funded, Nanosys Pursues Initial Product Launch by 2006," *Small Times,* July 21; Nanosys Inc. (2003–2005), Press Releases.

8. Nano-Network (2006), "Northeast Ohio Nanotech Executives Head to Washington DC to Discuss Policy Issues," Press Release, February 15, (Cleveland, Ohio); Cleveland.com (2004), "Good Times for Ohio's Small-Sized Industry," March 16; Nanovip.com (2004), "Secretary of U.S. Treasury John Snow Visits Nanofilm," July 13; Optiboard.com (2002), "Nanofilm Adds International Sales Staff, Expands Presence in Europe," September 23; InvestorIdeas.com (2004), "Large Crowds Expected for NANO Week: Hundreds of Scientists, Entrepreneurs and Executives to Attend," October 5; *Small Times* (2003), "Nanofilm Coating Makes Spectacle Overseas," November 26; Federal Reserve Bank of Cleveland (2005), "Business Advisory Council: List of Members," (Cleveland, Ohio); Neuman, S. (2004), "Nano-Tech in Northeast Ohio," Making Change Web, October 27; 90.3 WCPN and The Cleveland Plain Dealer (2004), "A Quiet Crisis: Manufacturing and the Greater Cleveland Economy: Transcript of Program," May 27 (Cleveland, Ohio); *Thomas Register* (www.thomasnet.com) (2006), "Nanofilm," Thomas Publishing Company; Nanofilm Website (www.nanofilmproducts.com) (2006), "About Nanofilm"; "Nanotechnology Firms Start Small in Building Big Future—Nanophase" (2000), *International Herald Tribune* (iht.com);
9. OE Magazine (oemagazine.com). (2005). "French Competitiveness Clusters Unveiled," July 25.
10. Discussion on clusters of the different countries come from a variety of sources, such as listed in References 9 and 12–15. In addition, refer to EU Commission (2003), *European Trend Chart in Innovation: Thematic Report and Cluster Policies* (Brussels. Belgium), pp. 7–26.
11. *Industrial Focus* (2005), "Dutch Nanotechnology Expertise," March/April.
12. Department of Trade and Industry (2001), *Business Clusters in the UK—A First Assessment: Vol. 3: Technical Annexes* (London, UK); Oxford City Council (2004), Oxford Local Plan 2001–2016, Topic Paper, "Employment in Oxford" (Oxford, U.K.); U.K. Minister of Science (1999), *Biotechnology Clusters.*
13. Oxford City Council (2004), Oxford Local Plan 2001–2016, Topic Paper, "Employment in Oxford" (Oxford, U.K.), p. 16.
14. Ibid., p. 9.
15. U.K. Minister of Science (1999), *Biotechnology Clusters,* pp. 3, 35.
16. For further discussion on these issues, refer to U.K. Minister of Science (1999), *Biotechnology Clusters,* pp. 24–43. A flurry of reports sponsored by the European Union from 2000 to 2007 support the 1999 report on these issues, indicating that little headway has taken place in Europe through the first decade of the twenty-first century.
17. Table constructed from information in "U.S. Cities Lead Knowledge Economy" (2004), World Business Section, CNN.com (April 15).
18. E.U. Commission (2002), *Improving Institutions for the Transfer of Technology from*

Science to Enterprises: Expert Group Report: Conclusions and Recommendations (Brussels, Belgium). For discussion of TTIs in general see pp. 8–10; for TTIs and their role in clusters, see pp. 23–24; for size and performance of TTIs in Europe and the United States, see pp. 26–30.

19. Ibid., p. 28.
20. Ibid., p. 30.
21. Yli-Renko, H., Autio, E., and Sapienza, H. (2001), "Social Capital, Knowledge Acquisition, and Knowledge Exploitation in Young Technology-Based Firms," *Strategic Management Journal,* Vol. 22, No. 6/7, p. 607.
22. See Reference 21 (above), pp. 587–613 for evidence of the growing role of small entrepreneurial firms in advanced technology clusters. Such studies emphasize the need for high-technology start-ups to access multidisciplinary knowledge bases that exist within clusters. See also Almeida, P. and Kogut, B. (1999), "Localization of Knowledge and Mobility of Engineers in Regional Networks," *Management Science,* Vol. 45, No. 7, pp. 905–917. Almeida and Kogut suggest the primary importance of entrepreneurial start-up firms in growing, technologically advanced regional clusters. Contrariwise, regional networks that come to be dominated by a few large corporations—such as occurred to the Route 128 (Massachusetts) Cluster—begin to stagnate. For another, earlier study on this very point see Saxenian, A. (1994), *Regional Advantage,* Cambridge, MA, Harvard University Press.

Chapter 10

Gatekeepers and Creative Clusters

GATEKEEPERS AND ADVANCED MATERIALS

A basic problem facing European attempts to compete with the United States hinges on the advanced clusters that increasingly are the source of innovation and productivity. As we have seen, European clusters tend to be extremely narrow in focus, such as centering on one industry and on academic science, and yet also highly fragmented, thus dissipating important economies and learning curve effects that technology clustering at its most useful can promote. The great advantage the American environment offers the innovative process in the late twentieth and early twenty-first centuries is the formation of complex, multidimensional cluster networks that, at the same time, exhibit singular cohesiveness, allowing the broad range of actors in the group to interact in close, symbiotic association and as an integrated, harmonized unit.

But this insight only begs the question of how do these multifarious advanced-material clusters form and become cohesive, integrated groupings? Can we assume that this process happens automatically, as, in the physical world, a crystal forms around a "seed" particle. This would imply that clusters form through some kind of geographical and technological determinism. For instance, a common belief is that early twentieth century advanced-material clusters that formed around a region like Niagara Falls and the Gulf of Mexico did so because of the existence of critically important natural resources—electric power and fossil fuels, respectively. However, upon closer study, the evidence does not support this rather simplistic model of industrial growth, not in those cases, and certainly not as the U.S. advanced-materials industry approached the twenty-first century. There is nothing predetermined about the process. Rather, the human element as it operates within complex social networks must be carefully integrated into our understanding of cluster siting, formation, and expansion. This question is explored in more detail in this chapter. In doing so, we learn more of the mechanism behind evolving clusters and, thus, probe why certain clusters succeed while others, such as those in Europe, struggle, stagnate, and fall behind as global competitors.

The Advanced Materials Revolution. By Sanford L. Moskowitz

At the heart of the greater number of successful clusters—successful in the sense that they are larger, more diverse, more deeply creative, and commercially far reaching—within the United States compared to other parts of the world are the active and wide ranging activities of certain pivotal individuals known as "gatekeepers." These gatekeepers might be viewed as antispecialists in the sense that their background, education, interests, and motivations compel them to operate within, and travel between, different environments: scientific, technical, business, financial, political, and so forth. The importance of these gatekeepers is in their ability to push ahead the most important of society's innovations. In regions, and even some clusters, where gatekeepers may play little or no role, radical technological change rarely takes place. This is because truly pioneering innovations require strict coordination and control of the different components—technical, economic, market, and political—of the innovation process. In contrast, nongatekeeper clusters typically are of older technology and are characterized by firms that merely improve their traditional manufacturing skills. This pivotal coordinating role of gatekeepers rests on their ability to inhabit different worlds simultaneously and, therefore, quickly and efficiently transfer, direct, control, and synthesize those diverse strands of information, knowledge, capital, and skills that are crucial to the creation, commercialization, and diffusion of the most novel and forward-looking technologies and systems.

These individuals proved critical to the rise of modern American technology and the economic growth it fueled. The histories of American organizations and technology typically concentrate on the decades prior to World War II. This focus on the early twentieth century implies the emergence after the war of technical innovation centered in the large and continuously operating research and development departments of major multinational corporations. These departments reflected the rampant professionalism that entered into American businesses as they evolved into compartmentalized organizations unto themselves, inhabited by growing battalions of nameless, faceless specialists. In such a standardized, "lock-step" world of many, relatively obscure personnel working within ever narrower lines of research, whatever creativity might occur, mostly of an incremental variety to be sure, would at least continue to push the bounds of efficiency bit by bit. This controlled, less-than adventurous type of creation, far more than the radical and upstart "Eureka" moments generated by the unpredictable "genius," was far better suited to an age that hailed the rise of the "organization man" and the rigid (and comfortable) conformity that he represented.

But has the modern world truly ousted the individual genius from his perch and marginalized him as irrelevant to society? In fact, since 1980, we notice a resurgence on the technology stage of the gatekeeper. In describing a successful manager of technology in the twenty-first century, that singular person possesses a rare combination of qualities that we have suggested above: A technical competence, an in-depth understanding of academic culture, and market and business acumen, combined with a protean grasp of managing multiple networks and complex relationships.

The reason for this resurfacing of this veritable "Renaissance man" in advanced-technology creation may have to do with the importance of the small firm as the center for such innovation, and the need for these firms, not being by their very nature integrated or diversified, to group together in clusters to achieve necessary economies. The gatekeeper must take center stage once again as the great collector, processor, and synthesizer of information from essential and divergent environments that coexist within a region. Most often, the gatekeeper founds a central or flagship firm, most often a start-up, around which other firms within and surrounding the industry congregate and link together. The gatekeeper today keeps order in a cluster, ties firms within the cluster together, and makes sure information skills, and technologies flow freely between the encompassed firms. If technology, economic growth, and productivity today arise by integrating a number of environments, it is the gatekeeper who is reinvigorated as the very lynchpin of this process.

AMERICAN STYLES OF GATEKEEPING AND ADVANCED-MATERIAL CLUSTERS

We saw in the previous chapter the impressive breadth and depth of America's high-technology cluster environment. The richness of the American gatekeeping tradition is evident today as well. No other country matches the United States in terms of the diversity, range, and creativity of their gatekeepers and their active pursuit of new technology and markets, both within the United States and internationally. The following sections examine the types, activities, and achievements of American gatekeepers working today within the U.S. advanced-material industry. In doing so, we identify three distinct types of gatekeepers that have been, and continue to be, central players in the rise and expansion of America's advanced-materials revolution. Our case studies examine the distinct backgrounds, interests, and skills of these three types of entrepreneur–gatekeepers. Despite these differences, these cases also suggest their common roles as the source of successful companies anchoring regional, multienvironment cluster groups that are the source for new-generation global advanced-material technologies. The case studies examined involve gatekeepers working within a large and expanding cluster group that we have previously discussed (the Albany area of New York State) as well as gatekeepers associated with developing relatively new cluster communities (Massachusetts and Ohio).

Following these descriptions, we compare the U.S. situation with Europe's, specifically the United Kingdom. Particular case studies illustrate the general point that the shortcomings of European clusters vis-à-vis the United States can be traced to the dearth of variety, range, and "multidimensionality" of European gatekeepers. For example, we observe that European entrepreneurs adhere too close to academia and its narrow, theoretical scientific traditions. It is in such matters that competitive advantages and disadvantages arise. At the same time, the United States cannot expect its competitive lead to remain unchallenged, especially in such a dynamic environment as that of the European Union. Indeed, we conclude the chapter by providing some evidence Europe may be beginning to adopt the American "gatekeeping" style of innovation, with successful outcomes for firms and the clusters they inhabit.

Investor Gatekeepers [1]

We have noted the critical importance of the venture capital community to American success in the development and application of the new generation of advanced materials. We tracked the tasks performed by this community that are essential for the proper functioning of the seamless web of creation that occurs within and between advanced-technology clusters. It is not surprising, then, that those we might term "stars" of the profession are important agents in the creation and expansion of advanced-materials cluster groupings. These "investor gatekeepers" move comfortably from the world of engineering to the realm of business and finance. Often educated in both fields, their work-a-day experience over a period of years certainly imparts flexibility of function and interests to these pivotal actors. They may wield their influence within a single cluster region or, through their financial contacts, extend their activity over a number of clusters covering a wide geographical area. We need only consider the career trajectory of the high-technology venture capitalist Steve Jurvetson as an illustration of this type of gatekeeper. As a result of his own success as a highly active investor–gatekeeper, Jurvetson has been honored in the *New York Times, Forbes,* and other publications as one of the ten most influential venture capitalists in the world.

Born, raised and educated on the West Coast, Steve Jurvetson has left his imprint on quite a number of important companies and promising industrial clusters. As do many inhabitants of Silicon Valley, he began his career in technology. He graduated with a Bachelors and Masters of Science in electrical engineering from Stanford University. Soon thereafter, he combined science and technology with business by adding an MBA from Stanford to his academic portfolio. Jurvetson's professional experiences reflected his broad-based education as he moved from pure engineering to business and market management. Jurvetson's first professional position was at Hewlett Packard (HP) where he performed R&D duties as a programmer and a materials science researcher. While at HP, he patented a number of his communications chip designs. Jurvetson then distanced himself from pure research and design to do product marketing and placement for Apple and NeXT Software. After a few years, he became a consultant to Bain & Company, where he developed and implemented marketing, sales, and business strategies for a range of companies in the software, networking, and semiconductor industries.

At this critical juncture in his career, Jurvetson took what he knew from engineering, business, marketing, and consulting, and parlayed that accumulated knowledge and expertise into high-technology venture capital investment, thus adding financial services for start-ups to his roster of expertise. Jurvetson's first significant foray into the venture capital field found him helping finance the then-new service Hotmail. His background in engineering, product design, and market development within the IT field proved invaluable to him in assessing the benefits and potential problems with this move and pinpointing an appropriate amount of capital to invest. Having helped establish Hotmail, he sold his investment to Microsoft for more than $400 million.

The field of advanced materials and their role in IT and communications technology appeared to him to be an attractive path to pursue, and one that played well into his multifarious knowledge base and talents. As an investor, first and foremost, Jurvetson's critical eye ranged widely over technology and geography in search of the most promising companies to back. By 2004, Jurvetson's firm, Draper Fisher Jurvetson, had invested in sixteen major advanced-materials and nanotechnology firms, more than any other venture capital firm in the United States or internationally. These firms, operating in clusters throughout the United States (and Canada), included Arryx Inc., BinOptics Corp., Coatue Corp., D-Wave Systems Inc., Egeen Inc., Imago Scientific Instruments Corp., Konarka Technologies Inc., Luminus Devices, Microfabrica Inc., NanoCoolers Inc., NanoOpto Corp., Nantero Inc., NeoPhotonics, SiWave Inc., Solicore Inc., and ZettaCore.

Jurvetson's approach to investment has been, as expected, broadly based and, in essence, contextual. Investor–gatekeepers on the order of Jurvetson are in a unique position in the field of advanced materials. They provide considerably more than a conduit for investment money into these regions; they identify and define, out of all possible investment decisions, those most promising technologies that stand the greatest chance of successful development and market entry. They also understand that it is vital for their interests that they back those firms that look to become major participants in a growing high-technology cluster. Such clusters provide them with synergies as well. These regions, because of the technical, economic, and even political linkages and the resulting agglomeration benefits that that emerge within them, offer a reliable and (relatively) risk-free location for expanding venture capital placement. Table 10.1 shows the companies within which Jurvetson's firm has invested (as of 2004). In virtually every case, these client companies operate within

Table 10.1. Draper Fisher Jurvetson investments as of 2004: company and location

Company	City	State or province	Country
Arryx Inc.	Chicago	Illinois	United States
BinOptics Corp.	Ithaca	New York	United States
Coatue Corp.	Woburn	Massachusetts	United States
D-Wave Systems Inc.	Vancouver	British Columbia	Canada
Egeen Inc.	San Francisco	California	United States
Imago Scientific Instuments Corp,	Madison	Wisconsin	United States
Konarka Technologies Inc.	Lowell	Massachusetts	United States
Luminus Devices	Woburn	Massachusetts	United States
Microfabrica Inc.	Van Nuys	California	United States
NanoCooolers Inc.	Austin	Texas	United States
NanoOpto Corp.	Somerset	New Jersey	United States
Nantero Inc.	Woburn	Massachusetts	United States
NeoPhotonics	San Jose	California	United States
SiWave Inc.	Arcadia	California	United States
Solicore Inc.	Lakeland	Florida	United States
ZettaCore	Englewood	Colorado	United States

Source: [2].

a growing high-technology industrial environment. In a number of these cases, the advanced material company is a central player in this cluster formation and expansion. Arguably, Jurvetson's greatest influence is in his financial, technical, and managerial support of important companies within nascent but promising cluster groups. Such is the case, for example, with Imago Scientific Instruments of Madison, Wisconsin.

Investor–gatekepers like Jurvetson make it their business to embed themselves within early-phase clusters that their investments help to spawn and nourish. Their business opportunities expand directly as the cluster in which they have invested evolves. Because of their strong backgrounds in science, technology, and engineering on the one hand, and business, finance, and marketing on the other, these investors move freely from one company to the next in a cluster, and from one field to another within a company. They know what investors want in a business plan and financial and managerial structuring, and they understand the technology behind the businesses, which ones are likely to succeed and which ones not, how they should be managed and financed over their product lifecyles, and so forth. They network easily between different actors. Because they speak the language of business and of technology, they, more than any other participant, can link together the disparate elements within a cluster group, and strategically place seed money in those additional businesses that add organically to the future success of the technology complex as a whole.

It is for these reasons that investor–gatekeepers play an important part in the future development of companies and clusters in which they themselves have invested. Jurvetson himself has been very active in state and regional groups and alliances that support advanced-materials technologies and their market applications. Most notably, he served as cochair of the Nano-Business Alliance, the main lobbying and support organization for U.S. nanotechnology business. Overall, the most successful of the investor–gatekeepers wield enormous influence on twenty-first century high technology and the regions that spawn them.

Engineer Gatekeepers

Although gatekeepers like Jurvetson were trained as engineers, this training and expertise serves as background, albeit very important, to their primary career: finance. There are other types of gatekeepers, also extremely influential in their fields and within the cluster or clusters in which they operate, who are both by training and profession engineers. At the same time, they have learned and readily handle and coordinate the business, financial, and political aspects of their business. It is important to note that, just as most financially oriented professionals do not have technical or political aptitude and, therefore, cannot be called gatekeepers, so the majority of trained engineers, even those involved in start-up firms, do not comfortably span the business, financial, and political worlds. Although these people may be superb performers in their chosen profession, they have limited impact individually on the technological change and economic growth of a country.

Such individual influence generally resides in those true gatekeepers who, engineers though they may be, by dint of their training, aptitude, and interests, operate comfortably beyond the engineering field and within broader, multidiscipline environments.

In an earlier time in U.S. technological history, movers and shakers in new materials industries such as Frederick Beckett of Union Carbide in the Niagara region, were model engineer–gatekeepers (sometimes referred to as "engineer–entrepreneurs"). In the early twenty-first century, innovators and entrepreneurs such as Clinton Ballinger, Davis Carnahan, and Scott Rickert represent a more modern paradigm of the engineer–entrepreneur working and extending his reach within a well-defined high-technology cluster. These examples show how devoted engineers absorbed, oftentimes out of necessity, into their skill sets an understanding of financial leverage, business strategy, and political maneuvering, and how they synthesized these various abilities, whether native or acquired, into a coherent strategic plan for technology transfer, firm growth, and cluster expansion.

Clinton Ballinger was one of the important gatekeepers working within, and helping to form and direct, the burgeoning advanced-material industrial group in upper New York State [3]. With advanced degrees in electrical and nuclear engineering, Ballinger worked a number of years at Lockheed Martin in New York State, in the Advanced Concepts Research Division, on diverse scientific and technological projects. While at Lockheed, he published and patented prodigiously in such areas as optoelectronics, computational physics, nuclear science, and medical physics. From this background, Ballinger, along with two colleagues, in 2000 left Lockheed to start their own company (Evident Technologies) that exploited their knowledge and experience in optoelectronics as applied to biotechnology, in particular, the development of biochips for clinical applications.

At this point, Ballinger learned the ropes in New York State investing. He began to make contacts with both the political and the investment communities in the region. Given the highly political nature of cluster building in the state, Ballinger's ability to span the technical, political, and financial arenas proved absolutely critical to his success. From this networking, and the development of convincing business plans, Ballinger managed to close on $3.8 million of seed money from various and strategic sources in the region. From this beginning and its expanding political and financial contacts, Evident Technologies under Ballinger held a prime place as one of the real landmark companies, and a central organizing force, within Albany's growing advanced-material cluster complex.

Another example of the engineer entrepreneur is Davis Carnahan, President of Nanolab, Inc., located in Massachusetts and developer of advanced drugs, computer chips, and optical components [4]. Carnahan obtained his Master's Degree in Ceramics Engineering and worked for years in Massachusetts industry in such technical fields as composite materials, processing additives, and metal coatings. His technical skills catapulted new materials to market. He worked for W.R. Grace as a researcher, helping to develop new casting technology for thermal shock-resistant automotive engine components. While at Grace, he developed electrophoretic deposition for the production of optical fibers and invented a process to create carbon composite valve lifters, which are now used in Briggs and Stratton engines.

But his interests extended well beyond the scientific and technical to business and marketing. After his stint with Grace, he worked for smaller start-ups and so gained much experience with small- to medium-sized firms. He also came to understand the "business" side of innovation and patenting. His technical, administrative, and business acumen and expertise touched on many industries, including mechanical and automotive, ballistics and defense, refining, and advanced materials in general. He served as engineer at CeraNova, developing and guiding to market advanced-material products such as high-temperature superconductor components, piezoelectric actuators and sensors, and polymer–ceramic composites for ballistics systems. He then moved on to Busek Co., transferring his knowledge from CeraNova in the development of silicon carbide components for use in the ethylene cracking industry, and directed programs in ion-beam synthesis of microelectromechanical systems (MEMS), plasma synthesis of carbon nanotubes, diamond-film deposition, and field emission. In his work at Busek and CeraNova, he created contacts in the business community and, most importantly, with investors, especially angel investors, and the trade media for the high-technology industries.

In 2000, Carnahan went off on his own and started NanoLab, a start-up company that licensed an emerging technology from scientists at Boston College involving scientific work in the nanotube field. Carnahan's business network, established in his work for Grace, CeraNova, and Busek, served him well in obtaining seed money for early operations. These New England-based contacts, which extended through Massachusetts and Connecticut, increased the exposure of and tapped funding for the company. Boston College was a crucial link to clients through its International conferences on nanotube technology. Eventually, Carnahan and his company captured Massachusetts-based companies as clients, including Raytheon. Carnahan's firm, in a similar manner as Evident Technologies in New York State, now stands at the center of a growing industrial cluster in Massachusetts as it positions itself to become a leading supplier of advanced nanotubes to the flat-panel-display industry.

A third instance of the engineer-gatekeeper is Scott Rickert, founder and president of Nanofilm Inc. in Cleveland, Ohio [5]. Described as the tenth-best nanotech industry nationwide, the region around Cleveland is a growing nanotechnology area. Richert and his company have been at the center of action in this region's growth. Rickert, like the other gatekeepers, started his career with a technical background. He obtained his doctorate in chemical engineering from Case Western Reserve University. While at Case Western, he conducted research into new polymer materials. He then entered the entrepreneurial arena by transferring his work from the university into his start-up company dedicated to the development of thin films. Rickert proved himself no narrow academic. He absorbed business thinking and strategy, especially from firms he dealt with in one way or another in and around Cleveland. Out of his business experience, he distilled and applied a set of practical business approaches as a start-up firm evolves. He advocates that in financing the start up, it is essential to rely more on self-financing, state loans, customer financing, and vendor financing, and less on angel investors. He supports the limited-liability strategy as optimal for start-ups since the founders can build their basis in their equities over time and can directly benefit from losses generated in the first several years. As the firm enters its growth phase, he supports strategic partnerships and joint development agreements with major mainstream OEMs.

Prior to the establishment of Nanofilm, only a fragmented, nascent cluster existed that did not allow synergies to come into play, as they had, for example, in Silicon Valley. Rickert himself described the problems facing Ohio as a potential center of growth:

> Part of the problem [stemmed] from the parochial nature of the region. Researchers at Case Western Reserve University [didn't] always communicate with each other, much less their biomedical peers down the street at the Cleveland Clinic. . . . Case only recently convened everyone at the university who claims they work in the nanofield. . . . Imagine the difficulties of figuring out who at the University of Akron is working on something that neatly fits in with the needs or research of someone in Cleveland or Kent. . . . The whole point of having a business cluster is to create competition and creativity; to foster entrepreneurial spin-offs; and to lure venture capital to pour money into these new businesses. [6]

Rickert carried the idea of the importance of advanced-material clusters as coherent regions of interlinkages, controlled communications, and directed knowledge transfer to capture agglomeration economies to its logical and ultimate extent :

> I think we need to first network . . . in the nanocommunity here . . . we are just starting to get organized. I can't believe how many nanotechnologly companies there are in this area. . . . You never knew them because there was no way to network . . . and there are all sorts of ones involving ceramics to metals, plastics. . . . We all have that common bond and we're trying to build an industry and we're starting to network, starting to work together to partner. . . . [For example,] Ferrell is a big nano company now and they are just down the street from us. I didn't know that. Now I do. We're starting to work together. [7]

Rickert, the prototypical twenty-first century engineer–entrepreneur, appears to understand all too well the importance of forming an interlinked high-technology cluster community and the multidimensional benefits that arise from that. He also knows instinctively that his unique set of skills handed him the opportunity to spearhead such an undertaking. Rickert's strategy has been to integrate this cluster and become the center of its growth, employing workable business strategies. He initially tapped family and friends around the Cleveland area for seed money. He then began establishing business relationships with companies around Cleveland, including Goodyear, which supplied additional seed funding. This regional networking led to an eventual entrée into LensCrafters, the company's first important customer. He then linked into an expanding battalion of business, financial, academic, and government organizations and agencies, including Kent State University, Ohio State University, Wright Patterson Air Force Base, the Glenn Research Center, the Cleveland Clinic Foundation, and the Ohio Business Technology Center.

As Rickert became known within the business community, he participated frequently in conferences and industry efforts to obtain government funding for high-technology clusters. He helped form and closely allied himself with Nano-Network, the Ohio-based organization composed of scientists, entrepreneurs, and financiers and dedicated to improving and expanding high-technology clusters and research and commercialization activities and capacities in the Northeast Ohio area. Rickert also represented the area's only advanced-technology company on Ohio's Business Advisory Council, which advised the Federal Reserve Bank of Cleveland on business conditions useful in making monetary policy decisions. These organizations, with Rickert and Nanofilm as predominant components, supplied the connecting link that coordinated knowledge, information, and technical transfer within and between companies, institutions, and agencies within this growing Ohio cluster.

Corporate Gatekeepers

Corporate gatekeepers come to the advanced-materials field with their core experience in the business world. They are true entrepreneurs in the sense that they often have owned and operated firms within a region. These initial companies generally touched on the materials field in one way or another, such as chemicals or petroleum products. As is true of the financial and engineering gatekeepers, these gatekeepers also began their career in the sciences or technical fields but soon after graduating with their degrees, and with a possible short stint in technical work, gravitated toward exploring corporate and business opportunities. In common with the other types of gatekeepers we have examined, corporate gatekeepers also know intimately and travel readily within and between the major components of the multidimensional environment that defines the modern, high-technology cluster. Like those others, these personalities play the part of active agent understanding the need for interconnectedness between firms, disciplines, and fields within a regional complex. Through their broad understanding of different endeavors of activities, but with their corporate base as, as it were, their home key, they inhabit with ease and flexibility, and bring together into a comprehensive and integrated strategic whole, the different realms—economic, business, technical, and political—that convert technological possibility into economic reality. The following case examples illustrate the activities of the corporate gatekeepers within three separate real-world contexts.

CNI Inc. (Dallas Cluster, Texas) [8]

The story of CNI in Texas is an excellent case to illustrate the importance of a corporate gatekeeper in the rise of a significant advanced-materials company and of the cluster group he helped create. One of the major players in the story of carbon-based materials known as fullerenes, CNI Inc. supplies fullerenes to nearly 500 customers worldwide. CNI holds over 100 patents in the field related to the production of the material and its applications for sensors, electronics, computing, lightweight materials, and drug delivery.

Richard Smalley, the scientist who helped pioneer the field, is rightfully described as an important part of the CNI story. He certainly played a leading role in the formation of the company, a start-up based on his own research and patents. While remaining at Rice University, he continued to guide the scientific course of the company until his death in 2005. He also was responsible for coaxing fellow chemist Robert Gower to join CNI. Gower, it turned out, was the one who played the pivotal gatekeeping function for the company and the cluster which is growing around it.

Gower, a Ph.D. chemist and former researcher, embraced business and finance early in his career. Over the decades, Gower proved himself a highly capable businessman. His ability to work within the technical, business, and financial worlds afforded him the opportunity to become a superb manager of petrochemical companies. He initially moved into operational and managerial positions for companies like Sinclair and Atlantic Richfield. He parlayed this experience and his growing network to take the helm as CEO of Lyondell Oil, a company he helped to turn around financially in the 1980s.

By the 1990s, Gower had amassed a track record of turning money-losing petrochemical companies in Texas into healthy billion dollar companies. Looking for new opportunities, and at the urging of Smalley, Gower agreed to take the reins of CNI. He provided his managerial expertise, and over $1.3 million of his own money, to the company. From his experience, Gower understood that, even if initially emerging from academic research (as was the case with CNI), a technology company can succeed only if it ultimately distances itself from the unworldly bias of universities since "universities . . . are generally not skilled at converting [their own disruptive technology] into . . . something which can be used by industry to benefit society" [9]. Gower's management background led him to make sure that academics remained outside of the upper managerial levels of the company. He recruited experienced colleagues from past associations to fill key positions such as CFO.

Gower also made use of his extensive contacts in the business community. He helped to secure laboratory and pilot plant space at a facility run by the industrial engineering and construction firm Kellogg Brown & Root. He also secured $15 million in venture money from local investors. Looking outside of the immediate cluster, Gower directed the company toward potential customers with whom he had dealt with in the past, such as IBM, Sumitomo, and DuPont. In 2002, he shepherded an agreement with DuPont Central Research and Development (CR&D) to license CNI's patented laser-oven buckytube production process for use in the area of field-emission flat-panel displays. And again, in 2004, CNI entered into a joint development agreement with manufacturing operations in the United States, Germany, Japan, Malaysia, and Singapore to develop advanced polymer products using single-wall carbon nanotubes.

These global arrangements spurred further growth of CNI's Texas cluster as these and other multinational companies sent researchers and managers into Texas as part of mutual R&D and executive training agreements. Gower called upon his network of contacts to bring funding from the state government into the region. He was one of the founders of the Texas Nanotech Initiative (TNI), a state-wide consortium of Texas-based universities, industry leaders, investors, and government officials that pooled their collective reputations and resources on generating funding for Texas' advanced-technology companies. TNI proved to be a catalyst in the creation in 2005 of the Texas Emerging Technology Fund (TETF), which dedicated substantial funding to promising advanced-material enterprises within Texas' growing high-technology clusters.

Nanosys (Silicon Valley, California) [10]

Silicon Valley continues to spawn new-technology firms, a growing number of which involve advanced materials. The corporate gatekeeper plays an important role here. The case of Nanosys is instructive. As previously described, Nanosys, founded in 2001, develops nanotechnology systems composed of such materials as nanowires, nanotubes, and nanodots. These devices offer improved performance in speed, sensitivity, and power consumption.

At first blush, it appears as if the success of the company hinged on the efforts of a typical *engineer*–gatekeeper. Charles Liber, Harvard Professor and Nanosys founder, worked on the theory of nanoscale logic gates and on reproducible bottom-up assembly of basic logic design elements. He became the world's leading authority on the synthesis of nanowires. His laboratory created prototypes for nanotechnologies for electronic and optoelectronic applications. For his work, he won the Feynman Prize for Nanotechnology and the Leo H. Baekland Award from the American Chemical Society. Three members of his team at Nanosys were named as the world's top 100 Young Innovators by MIT Technology Review. Liber himself served as a Member of Board of Directors and Chairman of the company's Scientific Advisory Board.

However, these men were basically scientists and not well linked to the broader business and financial community within Silicon Valley. It was the CEO of the company, Larry Bock, who worked within and beyond Silicon Valley, spanning different areas of expertise, to support Nanosys' growth within the region. As did a number of corporate gatekeepers, Bock started out in science, with a BA in biochemistry from Bowdoin College. He then added to his business capability with an MBA from UCLA. As he proceeded in his career, Bock became known as a deft synthesizer of the technical and business worlds. As one colleague observed, Bock "is one of these people who looks at the research and the resources available and how they can put it all together" [11]. Bock's strategy, which served Nanosys well, reflects this synthesis in that it demands a coherent platform technology, the use of multiple corporate partners with which to leverage the platform, and a product portfolio to diversify risk.

Bock's career clearly links these various elements. While starting out as a researcher, he eventually moved into business management and entrepreneurship. Bock became cofounder of over a dozen companies and served as CEO of six of them. He became a well-known biotechnology entrepreneur, managing 14 start-ups in the field. Prior to joining Nanosys, Bock was General Partner at CW Group and Avalon Ventures, venture capital management firms specializing on seed-stage start-ups. Through this work, Bock was selected by *Venture Capital Journal* as one of the Ten Most Influential Venture Capitalists in the United States. By drawing on his scientific and business experiences, he made contacts with research groups in the area and secured the rights to intellectual property developed by UCLA, Berkeley, and Lawrence Berkeley National Labs. These became important inputs for Nanosys' competitive positioning. Over time, Bock began to exert political influence in the region. He leveraged this position to help fund and organize the Northern California Nanotechnology Initiative, an organization instrumental in tapping and directing state funds into the region to help develop technology and business there. To further the advanced-technology impetus of the region, in 2006, Bock's funding endowed a chair in nanotechnology at Berkeley. This endowment signaled that the region was a major area in advanced materials. It thus attracted further funding from local and state venture groups as well as additional companies who settled in to benefit from the area's growing economies.

Nanophase Technologies Corp. (Illinois) [12]

Nanophase Technologies in Illinois is another example of a company based on the work of a well-known scientist–engineer but that depended for its growth on a different and more wide-ranging sort of talent. The industrial cluster that has been forming around Nanophase, occupying a large and growing geographical area, is one of intricate and close ties between companies and organizations across the technological–financial market–political landscape. The evolution of Nanophase under the stewardship of Joseph Cross reflects this multidisciplinary approach to technology creation and market penetration.

The company was spun out of research work done by Richard Siegel based on techniques for making nanocrystalline materials. Siegel, a Professor of Materials Engineering at Rensellaer Polytechnic Institute (RPI), excelled as an engineer but did not play a major managerial function in the company. Cross, the President and CEO of Nanophase, embraced this role. He too began his career in science and engineering, with his B.S. in chemistry from Southwest Missouri University. But, as others whose careers we examined, he pursued his MBA rather than an advanced degree in science or engineering. His background in science, engineering, and business would serve him well in the guiding Nanophase into the market. By the time Cross became part of the company, he already had a background of successfully directing several high-technology start-ups, including telecommunications companies.

Cross' strength at Nanophase came from the fact that he combined a scientific–technical understanding with practical business strategy, and he clearly discerned the distinction [13]. He referred to the early-stage Nanophase as a "science shop" but pointed out the company's later need to be a real business focused on commercial applications. Cross believes that many start-ups today grossly underestimate the length of time it takes to commercialize a technology. He noted that

"What works well on a bench and what works well for an hourly employee in a manufacturing scenario is an entirely different thing [and accordingly it was necessary to] change our business model in 1999 from a science development company toward a business focused on markets and revenue growth" [14]. Accordingly, when Cross came to Nanophase in the 1990s, he "beefed up" the company by bringing in his network of experienced business people who also had scientific and engineering backgrounds. His management team not only spanned science, engineering, and business, it also brought into the company an important interindustry experience. Cross contended that growth depended on the company "having a solid management team with multiple start-up experiences in several different industries. . . . [The management team must be] focused on the business, not just the science" in order to accomplish business goals: reduce the variable costs of manufacture and grow the gross margin [15].

As the company matured, Cross practiced a business model that reflected the importance of a close dialogue between the major environments that fundamentally shaped technologies with economic potential. Cross' model emphasized the importance of continued R&D in the company but only in close association with the will of the market, since technology and the market need one another throughout the development, commercialization, and positioning phases. Indeed, he and his team secured Nanophase's status as one of the industry's early starters by focusing on partnerships, customer service, and market applications [16]. The strategy helped move the company toward profitability (until the recession of 2000–2002). This linkage sharply cut the time for commercialization of innovation and shortened the lead time to market. Cross also advocated partnering with a motivated market leader to achieve improved market penetration. Cross secured such partners as BASF (sunscreens and personal care products), Rohm & Haas (codeveloped a slurry for semiconducting polishing), and Altana Chemie AG (nanocomposites for coatings, paints, polymers, plastics, and sealants). In Cross' model, Nanophase undertook nanomaterial development, engineering, and manufacturing while a complementary partner (like BASF) took on its shoulders what it did best: brand management, marketing, sales, and distribution on a global basis.

IS EUROPE IN THE "GATEKEEPING" GAME?

American technology today arises and develops within a rich and varied cluster environment, organized and guided by what we term gatekeeper personalities. These entrepreneurs, while varied in interests and expertise, generally begin their careers in technology but soon establish strong ties to business, financial, and political networks. They span these various disciplines and activities in a directed way and in the service of new companies developing advanced products and processes, especially in the new-materials area. These gatekeepers operate at the center of early and pivotal companies within regional clusters. These companies receive vital information, capital, technology, and political support through the networks that these gatekeepers create and nurture. But this process operates in the other direction as well. The growing resources of these important companies feed into and support the expanding cluster. They, in effect, serve as the flagship enterprises of the cluster group. As they grow in importance, they attract technical, business, and even political input from outside sources that, over time, through personnel, capital, and technology transfers, diffuses outward to the other entities making up the cluster. Eventually, the cluster achieves a reputation that attracts additional companies and personnel into the area, leading to its further growth and development. A symbiotic relationship builds and grows between the flagship firm and the surrounding environment. Agglomeration economies then take hold. This dynamic depends most of all on the multifarious abilities and activities of these central personalities.

The European situation contrasts sharply with that in the United States. Europe does not develop high-technology clusters that have the "critical mass" allowing economies of information and resources that are so critical to internal expansion of creative groups. Fragmentation and discontinuity plague cluster formation in the United Kingdom, France, Germany, and other Western European countries. In their assessment of the European cluster environment, Robert Huggins Associates found that E.U. regional clusters continue to struggle to "bridge the knowledge gap that would enable it to compete with the United States" [17], and that even the important industrial clusters in Europe were still concentrated in just a few regions and particular industries, and were not developing as fast or robustly as those in the United States.

At the heart of the problem for Europe is the lack of entrepreneurial and, even more important, gatekeeping talent. The creative reach of European clusters falls short of the benchmark set by the United States. The E.U. Commission understands that the European Union faces such difficulties. In recent reports that have been cited in previous chapters, the E.U. Commission maintains that innovation is above all spurred by entrepreneurial action, aimed at creating value through the application of knowledge. The Commission regularly expresses its concern that too few Europeans want to become entrepreneurs and guide the most innovative SMEs within growing high-technology clusters. It is true that there are some "serial entrepreneurs" in Europe that provide management expertise and informal monitoring for start ups, but little in the way of consistent, broad-based entrepreneurial energy and creativity operates within the European context. We have seen that there are significant cultural and psychological barriers to entrepreneurs in Europe. Even Europe's consumers are less open to new products than those in the United States or Japan, as evidenced by the continued U.S.–E.U. trade controversy over genetically modified foods (GMOs) [18].

The following case studies underscore some of the salient differences in entrepreneurial behavior between the U.S. and European advanced-materials industries. The cases of the U.K. companies Polaron and Oxford Nanoscience are particularly illustrative. They show how the lack of a multidisciplinary gatekeeper prevents a firm and the cluster surrounding it from expanding. Yet there is also evidence that Europe may be beginning to adopt the U.S. gatekeeping model to positive effect. The following example compares a European firm, which embodies the "traditional" European model of clustering and innovation with a U.S. start-up in the same field and examines the consequences in terms of national competitive advantage. A second case study, that of Oxonica, concludes the chapter and shows the more positive results from a European start-up helmed by an American-style gatekeeper who embodies the multienvironment approach to innovation.

Europe's Traditional "Gatekeeping" Model: Oxford Nanoscience/Polaron PLC (UK) [19]

Polaron is a British-based holding company comprised of four technology business components. Its more important segment is Oxford Nanoscience Ltd., a manufacturer of three-dimensional atomic probes. In recent years, the British touted Oxford Nanoscience as the key enterprise, the central anchor, of the advanced-material cluster forming around Oxford University. However, Both Polaron and its subsidiary Oxford Nanoscience found themselves facing difficult competition from American companies. In truth, Oxford Nanoscience remained a small player in its field and had had a difficult time distinguishing itself in the face of more successful U.S. companies.

The leadership at the helm of the British companies helps explain their disappointing performance when they squared off against their American competition. Certainly, the academic credentials of the British executives at the helm of Polaron and Oxford Nanoscience during the first years of the companies' existence impress one as steeped in the scientific underpinnings of advanced materials. The head of Polaron, Dr. George Smith, is a professor of materials and head of the Department of Materials at Oxford University. A specialist in field emission and field ion microscopy, he remains a theoretician in the area of physical chemistry and an early champion of courses in theoretical metallurgy at Oxford. Smith remains strongly embedded within England's scientific community. He continues to teach and conduct research at Oxford. He is an active member of England's most prestigious scientific organizations, including the famous Royal Society and the Institute of Physics. Polaron itself emerged out of pioneering advanced-materials research Smith conducted in the late 1980s on three-dimensional atomic probe systems. Through Smith's research, Polaron and Oxford University retain close and active ties. Smith continues to conduct scientific research at Oxford and then transfer his findings to applications at Polaron.

Smith's focus on the scientific aspect of the field, while vital in terms of new theoretical research, also limits both Polaron and Oxford Nanoscience competitively. The latter remained a basically small-scale, specialized, scientific instrument firm. Smith cannot be considered an entrepreneur–gatekeeper in the fullest sense because he does not straddle, draw from, and link the critical environments needed for full-scale competitive technology. In contrast to the American gatekeepers we have observed, he is not closely connected to the financial–venture capital community, a community that is in any case relatively small compared to the venture capital network in the United States. He also has not established strong links with the business community forming throughout the region or with the local governments. His main focus continues to be relying on his scientific work to nourish the firm. This means that Smith and his company remain wedded to an academic, laboratory-centered mind-set that competes in the world by improving products incrementally through scientific advances.

Smith’s approach to business favors the creation of specialty instruments made in limited runs for a few particular markets at high per-unit prices. The description of their business that appears on their website (as of 2007) reflects the fundamentally scientific focus of the company. The company’s basic “output” is described as “atomic probe microanalysis” that undertakes the “ultimate resolution for chemical analyses of materials for materials characterization capability.” Oxford Instruments advertises (mostly for the European audience) that its product attracts research laboratories and is the subject of numerous scientific papers. The fact that their product is accepted within scientific-laboratory circles is an important selling point, or at least is believed to be such, within European society. It is no surprise, then, that the markets for these instruments have remained generally limited to university laboratories, research institutes, and, to a lesser extent, the research laboratories of certain manufacturers, such as in the steel industry. Scientific excellence and selected markets, rather than large-scale thinking, characterize such niche strategies.

This approach to advanced technology might be said to reflect an essentially “aristocratic” mind-set, one often seen by students of the history of science and technology as rooted in the fact that, historically, the upper strata of European society applauded the pure sciences while devaluing practical engineering as proper activity for the lowly tradesman. This more narrow focus, certainly compared to U.S. entrepreneurs, must severely restrict the formation of the multidimensional relationships that need to be formed within the cluster community and that force a much broader sense of the strategic management of innovation.

The American Gatekeeping Model: Imago Scientific Instruments

Imago Scientific Instruments, an American company in a closely related field would far outpaced Oxford Nanoscience as a competitor. Despite its name, the company has been managed more as a industrial technology enterprise rather than a scientific research laboratory specialist. The background and experience of Imago Scientific Instruments and its management helps to explain the differing fortunes of the British and American companies. Imago develops and manufactures advanced industrial optical devices, such as atomic-force microscopes. Imago began as a start-up company that licensed research conducted at the University of Wisconsin–Madison by the company’s founder, Tom Kelly. Kelly had some business experience and a limited range of contacts in and around the region, which was beginning to develop as an industrial cluster surrounding the university. Kelly was able to raise the funds to start the company and to help it move toward its first commercial ventures. Growth remained an issue for the company.

Without further contacts within the business and financial communities, both within Wisconsin and beyond, Kelly was forced to relinquish the CEO role to another person, Timothy J. Stultz, who could tap that network. Stultz filled the true role of entrepreneur–gatekeeper for Imago. Stultz earned his doctorate in material sciences from Stanford University. But, in contrast to his British counterpart at Polaron, Stultz had considerable financial and business experience prior to heading up Imago. He put in 20 years in executive management within high-technology and capital-equipment companies. He founded and managed Peak Systems, a leading supplier of advanced thermal processing equipment for the semiconductor's industry. His most important position prior to Imago was as Vice President and General Manager of the Veeco Instruments Metrology Group. In this capacity, Stultz brought to market the world's first fully automated atomic-force microscope for use by leading chipmakers, including Intel, IBM, and Motorola, to locate tiny defects and measure materials at the nanoscale. While at Veeco, Stultz applied the concept that development of new technology must be done in close communication with the critical markets. Stultz made good use of his extensive contacts to position the company well in terms of strategic acquisitions and partnerships (Veeco had over $100 million of cash on hand to take advantage of such opportunities when they surfaced). This growing network imposed on the firm the requirement that it produce its instruments efficiently and market broadly and internationally. Stultz, by dint of his multidimensional, one might say eclectic, background, understood well that science and technology must be applied to create products with wide demand globally. By the time Stultz left the company, Veeco controlled nearly three-quarters of the world market in high-end metrology systems, primarily aomic-force microscopes, and Stultz had greatly expanded his pool of contacts within the global technology, business, and financial community.

Stultz did not forget these lessons, nor did he believe they only applied to Veeco. With his new position at Imago, Stultz brought his experience and networks to bear. The company began tapping a growing number of the larger venture capital groups with which Stultz had worked in the past, including the firms Draper Fisher Jurvetson, Infineaon Ventures, and Portage Ventures. The Jurvetson firm, in particular, worked with Stultz to strengthen the foundation of the company and expand and bolster the industrial cluster growing around it. Beyond access to a deeper investment pool, Stultz brought to the company a more focused direction aimed at taking advantage of strategic partnerships and linking Imago to the practicality of finding and exploiting new markets. Stultz understood the importance for the company to break free of its university ties and reach out to other companies, such as Intel, as partners in research, development, and marketing. Stultz maintained that ". . . to go from a company with a breakthrough product to one that can fulfill orders and distribute on a global scale takes investment in people and systems. . . . . . . [T]hese technologies [that we develop] must be more than for academic research [in order to for us to continue growing] . . . [we have] to be solving problems if someone is going to spend $2 million on an instrument. . . ." Stultz, who clearly embodies the American sense of practicality in business at the expense of science and the academic, and the use of whatever you can draw from to succeed, pointed out that, ". . . while . . . a large part of our revenue goes into R&D, we consider ourselves a market-driven company . . . we started as an R&D organization, but then hired applications people that are more users of the tool rather than developers. . . . [These people] can expand the number of ways the tool can be applied to an industrial problem." By 2006, under Stultz Imago had expanded rapidly, both within the United States and internationally. International sales, for example, had grown 35% since 2003. The company's customers included such large international concerns as Intel, Seagate, Toshiba, and the Central Research Institute of the Electric Power Industry [20].

Imago Scientific Instruments versus Polaron/Oxford Nanoscience

In April 2006, Imago acquired Oxford Nanoscience from Polaron. Clearly, Imago wanted to control some of the excellent scientific resources of Oxford Nanoscience. Having grown far larger than Oxford Nanoscience, Imago could incorporate into its organization the smaller and struggling British firm. By this action, Imago continued to grow and penetrate into new markets globally. Polaron, on the other hand, shrunk significantly as an enterprise.

Imago has clearly become a force in the industry. It's acquisition of Oxford Nanoscience reflects Stultz's general strategic thinking and his ability to dominate the industry. As a typical entrepreneur–gatekeeper, he could move easily between, and absorb and synthesize, the various inputs he needed from the different environments—scientific, technical, market, business, and financial—critical for innovation and market expansion. In contrast, the founder and CEO of Polaron did not range very far from his academic roots. The company suffered as a result. Within the region around the University of Wisconsin–Madison, Imago has become an important player, central to the growth of the region. The company creates employment and is intricately "networked" with other, complementary firms, many of which are start-ups licensing research from the university. Initially, Imago relied extensively on the capital, personnel, and research from the area. For example, it licensed its central technology from the University and borrowed seed and development funding from local angel investors. As the company grew under Stultz, it then looked further outward for funding, such as working with Jurvetson's firm, and markets, and it partnered increasingly with international outfits. These external contacts meant that inputs, technical, financial, and so forth, from national and international sources filtered into Imago and then, over time, as Imago interacted with its neighboring companies, outward into other firms within the Wisconsin area. The result has been job creation, resource transfer benefits, capital infusions, technology transfer, and a general acceleration in the growth of the regional economy.

The Winds of Change? The American Model in Europe: Oxonica [21]

Can we then assume that Europe must continue to run behind the United States in high-technology development and the economic benefits it bestows on its creator? The case of the United Kingdom's Oxonica at the least suggests that certain European high-technology enterprises can potentially compete with their U.S. counterparts on the world stage. At the same time, this case must be considered an exception to the general situation in the European Union and, moreover, one that very much proves the rule we have been developing. This case at the least indicates that the "Oxford" cluster may indeed have its champions. There is evidence that the American approach to entrepreneurship, either by chance or design, may be taking hold in that region. The management of this one company reflects this "gatekeeping" spirit with results that indicate future promise for it and the region that it calls home.

Oxonica, similar to Polaron, is a firm that emerged out of research conducted at Oxford University. Oxonica, spun out of Oxford university in 1999, specializes in the fabrication and commercialization of nanoparticle materials with innovative chemical and physical properties. The company was founded on the intellectual property developed by Oxford Professors Peter Dobson and Gareth Wakefield. The company began its entrance into the market with products used in emission catalysts for the automotive industry and cosmetics for UV protection.

During its first couple of years, the company was initially in the hands of Oxford-based scientist–engineers and then, as the company faltered, financial specialists took the reins. The company continued to decline as a commercial entity. Although Oxford University continued to supply the results of its scientific and technical research in the form of licenses for new-technology development, Oxonica lacked the guidance to assess and commercialize the most promising products and processes and prepare them for the marketplace. The scientist–engineers did not have the practical experience to commercialize these technologies, but the purely financial specialists who took over for the engineers lacked the technical expertise and intimate knowledge of technology markets to pick from the mass of possibilities the most commercially valuable intellectual properties.

With the company in financial trouble, and unable to penetrate into new markets, especially internationally, the entrepreneur Kevin Matthews was brought in by the company's Board to impose a broad-based approach to its corporate strategy. Mathews conforms very closely to the American model of the entrepreneur–gatekeeper. He received his doctorate in organic chemistry at Oxford University, followed by capturing a postdoctoral position that allowed him to explore the commercial possibilities of new surface technologies, pharmaceutical intermediates, and catalysts. But Mathews, while a product of academic science, did not remain at Oxford; he leveraged his talents and knowledge into a business career. Specifically, he looked for ways to find commercial pathways by which to bring new technologies into the market. In doing so, he obtained broad experience in technical management, financial analysis, business development, and international business practice. Matthews himself, clearly exhibiting the multidimensionality of experience characteristic of his American counterparts, describes his career as ". . . maintaining my interest in developing technology, and [being] responsible for starting new businesses by analyzing existing A&W businesses . . . [and in the process] gain[ing] expertise in restructuring businesses, taking products to market, and especially, understanding their technological development, marketing, and sales . . ." [22]. In the course of his work experience and prior to joining Oxonica, Matthews delved into technical development, business analysis, mergers and acquisitions, sales, marketing, and business strategy for companies such as ICI and Albright & Wilson. While at Albright & Wilson, for example, he designed and coordinated the "integration team" at the time when Albright was acquired by Rhodia in 1999. Following this acquisition, he undertook the position of global business director, managing a $100 million international operation with patenting and licensing responsibility, and in charge of managing and allocating significant R&D monies for the company.

Matthews' background and experiences, rooted as they have been in balancing technical development and business growth, meant that he had to appreciate science and technology, as did the Oxford University-based founders, but only at the service of practical business achievement. He understood that this balance was the only way Oxonica could succeed. When he took over the company, he contended that

> It helped that I . . . was hired into the company so [I] didn't have the hang-ups of the founder. I was able to be dispassionate and think more about what was best for the company and how it could grow. . . . We had to focus on some commercial opportunities very quickly. We could not continue burning money as a research house, which still had a "university approach." [Therefore] my biggest challenge when I joined Oxonica was to stop people doing research and accept the need to look outside, and to be more inventive in application of the technology. . . . Our business model concentrates on marketing and technology, with integration forward through distribution or partnership deals. We focus on adding value—matching market needs to new technology." [23]

With his understanding of science and technology, Mathews has certainly been adept at seeking out the most interesting and workable technologies across a wide range of applications, and avoiding projects with long time lines and heavy capital requirements. This expertise, coupled with his business background and network of contacts, meant that he attracted capital from the venture capital community far more readily than his more specialized predecessors, especially in the intensely risk-averse post-September 11 environment. Matthews spearheaded significant due diligence to select venture capitalists, and looked for those who could act as trade partners, opening doors for new business and helping to commercialize products. From his contacts and experience, he could choose an optimum syndicate of investors from local, national, and international sources who worked with him to expand the capital and market base of the company.

This multidimensional strategy proved successful, certainly relative to other U.K. (and European) firms that reside within the advanced-material arena. This leadership did not go unnoticed. Matthews won the *Small Times* 2005 Best of Small Tech Business Leader of the year. In winning the award, Matthews let it be known that, whatever success has been his, he has to tip his hat to the leaders in the field, the ones whose model of entrepreneurship and innovation management he has been moving towards. Matthews understood that ". . . being recognized in the United

States, where the [commercial] development . . . is highly active, illustrates that Oxonica is beginning to establish itself on the global stage" [24].

It cannot be said that Oxonica can yet compete against the most successful of the U.S. advanced-materials companies. It continues to design only a limited number of products and for far smaller markets than is typical for American companies. As of 2008, Oxonica has yet to achieve the efficiencies and cost advantages of U.S. operations. The strong pull of the theoretical sciences vis-à-vis engineering restricts the company's options in engineering advance. In fact, unlike many U.S. start-ups, Oxonica designs and markets its products but leaves the details of manufacturing to its licensees. But the devil is in just such details and, by relieving itself of this critical responsibility, Oxonica gives up the possibility of critical linkages of integration that the more engineering-minded American companies, who work on improved manufacturing themselves, are so successful in exploiting for competitive advantage.

This being said, Oxonica, with at least a portion of the American model incorporated into its culture, rather than Polaron, still holding onto an archaic, traditional European approach, has become one of the European Union's most important advanced-materials firms. It is the first nanotechnology company within the United Kingdom to be floated on the stock exchange, and has become a veritable hub and flagship enterprise at the center of Britain's fastest-growing high-technology clusters [25].

REFERENCES

1. Nanobusiness Alliance website (www.nanobusiness.org), (2005), "About the Alliance: Advisory Board—Steve T. Jurvetson; Web2.0 (www.web2con.com) (2005), "Steve Jurvetson: Speaker Biography (at web2.0 Conference); Draper Fisher Jurvetson Website (www.dfj.com) (2005), "Management Team."
2. Draper, Fisher, Jurvetson website (www.dfj.com).
3. Refer to sources on Evident Technologies in Chapter 8, Reference 5.
4. NanoLab website (www.nano-lab.com) (2005), "Nanolab Team"; NanoLab website (www.nano-lab.com) (2001–2004), Annual Reports; Kelly, M. (2004), "Nanolab Knows that Profits are Not Made on Nanotubes Alone," *Small Times,* March 9.
5. Feder, B. J. (2004), "Bashful vs. Brash in the New Field of Nanotech," *New York Times,* March 15; Nanobusiness Alliance website (www.nanobusiness.org) (2005). "About the Alliance: Business Leaders—Dr. Scott E. Rickert; Rickert, Scott E. (2006), "Taking the Nanopusle—Leadership in Nanotechnology: Our Nation's Leaders Recognize What's At Stake," *Industry Week,* March 8; Rickert, S. E. (2006), "Taking the Nanopulse—With Nanotechnology, Less is More," *Industry Week* (www.industryweek.com), January 18; Murdock, S. (2006), "Interview with Scott Rickert of Nanofilm", *Nanobusiness News* (www.nanobusiness.org) (2006); Nanovip.com (2005), "Nanofilm's Rickert Addresses Nanotechnology Partnerships in Industry Week," November 8; The Maple Fund website (www.maplefund.com) (2005), "Overview" (Scott Rickert); NanoTechnology

(www.nanotechnology.com). (2005), "Nanofilm's Rickert Addresses Nanotechnology Partnerships in *Industry Week,*" August 11; Investor Ideas (www.investorideas.com) (2005). "High IPO Hurdles for Nanotechs (Nanofilm)," February 23; Shryock, T. (2005), "Market Pull: Scott Rickert Has Led Nanofiln to Profitability Where Others Have Failed," Smart Business Network Online (http://cleveland.sbnonline.com), July; U.S. Treasury (2004), "Secretary Snow to Visit Cleveland, Ohio . . . to Meet with Local Business (Nanofilm)," Press Release, Washington, DC, July 9.

6. Neuman, S. (2004), "Nanotech in Northeast Ohio," Making Change Online (Cleveland, Ohio: Case Western Reserve University), October 27.
7. Ibid.
8. Reisch, M. (2001). "Gordon Cain," *Chemical & Engineering News,* Vol. 79, No. 32, p. 25; University of Minnesota (2006), "Bob Gower: Investment in Small Invention Leads to Big Business," *Institute of Technology News,* Winter; Reisch, M. (2001), "Gordon Cain: Autobiography" *Chemical & Engineering News,* Vol. 79, No. 32, p. 25; *LSU News* (2002), "LSU Alumnus and Donor Gordon A. Cain Dies at 90," Baton Rouge, Louisiana, October 31; Chemical Heritage Foundation (2001), "Everybody Wins When Gordon Cain Gets Involved: Autobiography Released in Paperback," Press Release, Philadelphia, Pennsylvania, July 5; Chemical Heritage Foundation (2001), "Gordon Cain," Fact Sheet, Philadelphia, PA.; Amato, I. (2001), "The Soot That Could Change the World," *Fortune* (June 25).
9. It is true that it was "extremely important" for Gower to obtain his Ph.D. However, he makes clear this achievement, while it gave him personal confidence, did not play any sort of direct role in CNI's ability to ". . . make and sell the material." Refer to University of Minnesota, Institute of Technology website (www.it.umn.edu) (2006), "Bob Gower: Investment in Small Invention Leads to Big Business," *News* (Winter).
10. Nanobusiness Alliance website (www.nanobusiness.org) (2005), "About the Alliance: Business Leaders—Larry Bock"; Lux Capital website (www.luxcapital.com) (2006), "The Lux Team: Larry Bock"; NanoBusiness Alliance website (www.nanobusiness.org) (2006), "Advisory Board: Larry Bock"; Nanotechwire.com (2004), "Nanosys Executive Chairman Larry Bock to Join FEI Company's Board of Directors," December 17; Nanotechwire.com (2005), "Nanotech Leaders Advance Policy Agenda on Capital Hill," February 9.
11. Forman, D. (2003), "Fully Funded, Nanosys Pursues Initial Product Launch, . . ." *Small Times* (July 21).
12. *The Wall Street Transcript* (2001), "Company Interview: Joseph Cross," Wall Street Transcript Corporation, New York, November 9; *The Wall Street Transcript* (2004), "Company Interview: Joseph Cross," Wall Street Transcript Corporation, New York, March 29; Cross, J. (2004), "Partnering Gives Smaller Nano Firms Access to More Markets," Small Times, November 4; Nanotechwire.com (2005), "Nanotech Leaders Advance Policy Agenda on Capital Hill," February 9.
13. The idea that Cross could span science, technology, and business as a prototypical gatekeeper is a clear theme that comes through in his interviews as well as comments from colleagues.
14. *The Wall Street Transcript* (2004), "Company Interview: Joseph Cross," Wall Street Transcript Corporation, New York, March 29.
15. Ibid.

16. Ibid. Also see Cross' own article in 2004 for *Small Times,* Cross, J. (2004), "Partnering Gives Smaller Nano Firms Access to More Markets," *Small Times,* November 4.
17. See Chapter 8, Reference 17.
18. See Sandblom, L. O. (2000), "Genetically Modified Organisms (GMOs): A Transatlantic Trade Dispute," Monterey Institute of International Studies.
19. Oxford Nanoscience website (www.oxfordnanoscience.com), Press Releases (Oxford, United Kingdom).
20. Gertzen, J. (2002), "Imago Lands $7 Million Investment," Milwaukee Journal Sentinel, September 25; Gallagher, K. (2005), "Big Leaps in Tiny Technology: Imago Builds Tools for Growing Industry of Nanotechnology, *Milwaukee Journal Sentinel.* June 4; Imago Scientific Instruments (2006), "Imago Acquires Oxford Nanoscience," Press Release, (Madison, Wisconsin), April 11; Imago Scientific Instruments website (www.imago.com) (2006), "Management Team," Madison, Wisconsin; Wisconsin Technology Council website (www.wisconsintechnologycouncil.com) (2005), "Imago Scientific Instruments Winner of First Governor's Small Business Technology Transfer Award," Newsroom, Madison, Wisconsin, June 7; *Manufacturing Business Technology* (2005), "Instruments of Innovation," September.
21. Oxonica website (2005), "Management Team," Oxford, United Kingdom; Hague, D. and Holmes, C. (2004), "Dr. Kevin Matthews, Oxonica," www.science-enterprise.ox.ac.uk, December; Oxonica website (www.oxonica.com) (2005), "Oxonica CEO Awarded *Small Times* 2005 Best of Small Tech Business Leader of the Year," Press Release, (Oxfordshire, United Kingdom), December 6; ADVFN News (www.advfn.com) (2005), "Oxonica plc: Acquisition of Nanoplex Technologies, Inc."; Wang, U. (2006), "Payoff Time for Nanotech," Red Herrings (www.redherring.com), May 22; *Nanoparticle News* (2003), "Matthews Sees Life Sciences and Energy/Environmental as Key," May 1; Matthews, K., (2005), "Oxonica Survives a Bleak 2001 by Adjusting its VC Strategy," *Small Times,* May 16.
22. Hague, D. and Holmes, C. (2004), "Dr. Kevin Matthews, Oxonica."
23. Ibid.
24. Oxonica (2005) "Oxonica CEO Awarded *Small Times* 2005 Business Leader of the Year," Press Release (Oxonica: Oxford, United Kingdom).
25. Oxonica may not be an isolated case of a slowly growing technological capability in Europe. For example, there is evidence that small, technology savvy start-up firms are on the rise in Europe. In fact, Reference 21 in Chapter 9, which we cited to support the contention that small start-up firms energize regional clusters, was a study that used survey data from 180 young technology-based firms in the United Kingdom that were involved in developing, commercializing, or manufacturing advanced technology in pharmaceuticals, medical equipment, information and communications technology, energy, and environmental products. See Yli-Renko, H., Autio, E., and Sapienza, H. J., "Social Capital, Knowledge Acquisition, and Knowledge Exploitation in Young Technology-Based Firms," *Strategic Management Journal,* Vol. 22, No. 6/7, p. 595.

Chapter 11

Conclusion: Broadening Horizons

In the previous chapters, we identified the key technologies that have come to the fore since the late 1970s, and those future "stars" that are likely to do so in the second and third decades of the twenty-first century. We identified which materials, markets, and regions look promising for current and future investment and which, although touted as the ones to watch, do not appear to manifest the type of growth model we would expect over the next few years. We have argued that the most important innovations flourish within the area that we call the advanced-material industry. We have shown that there is a coherent group of such products and their associated processes, that industry players understand that this group of innovations constitute a well-defined family of technologies, and that this industry lies at the very heart of technology change, productivity growth, and economic advance, and that these, in turn, determine which countries and regions rise (and fall) as safe havens for investment and, ultimately, become competitive forces in the world.

Our exploration of advanced materials, then, shows that it is not only an economically important industry but, taking this one step further, it is one of the most central to national competitive advantage. Accordingly, through the prism of the industry and its retainers, we address some of the most critical issues facing an increasingly globalized world. A fundamental question that we have asked ourselves in this narrative of the advanced-materials industry is why economic activity is spread so unevenly from country to country. We would expect quite the opposite in fact—a converging or "flattening" in economic progress as globalization expands. In our reading of the issue, such traditional explanations of divergence, such as differences in scientific ability, government spending on research and development, the number of large firms, and the availability of institutional capital, do not suffice.

This question, we have seen, focuses on the advanced materials industry as absolutely essential; within the industrialized nations at least, technical progress drives economic activity, and this industry is the prime representative for technological change in general. But what does this say about national and regional competitive advantage? Emphasis on globalization and its leveling force strongly implies that a nation or region cannot retain a competitive edge for long, that its advantage, whatever it may be, must eventually have to be shared with the rest of the

The Advanced Materials Revolution. By Sanford L. Moskowitz

world. In this sense, we can speak of "borderless" technology as knowledge and skill transfer from one country to the next.

But our understanding of technology creation based on the story of advanced materials in the United States and the European Union paints a very different picture. Local influences and biases, rooted in culture and history, as well as engineering and technical capability, critically determine the rate and direction of technological change. These forces do not so easily transfer from one place to another. More specifically, the degree to which what we have termed the "seamless web" of technology creation comes together into, and operates effectively as, an integrated, coherent system of actors and organizations varies considerably over space as well as time. We know that this happens through the subtle interplay and coordination of agents rooted in particular local environments. These entities—start-up firms, personnel, universities, state and local political persons, institutions, and groups—crystallize around central personalities known as gatekeepers. These agents of change either are themselves, or work closely with, regional venture capitalists who select those most viable technologies for development. Clusters then begin to form and expand outward from this center, and eventually link up and integrate with other, contiguous clusters.

The newest and most far-reaching advanced-materials technologies emerge not from general bold theories devised from above, by "big science" government, and academic science departments, but from small beginnings closely attuned to local economies and markets. These advanced and far-reaching discoveries find their way into the world from efforts to solve immediate problems, satisfy nearby demands, and bolster and enhance local economies through densely packed and closely linked industrial communities. As Anthony Venables reminds us,

> The key building block [of the new economic geography] is the recognition that proximity is good for productivity; dense configurations of economic activity work better than sparse or fragmented ones. Mobile factors—firms and possibly workers—will locate in order to take advantage of higher productivity, and this increases a positive feedback. Firms and workers go where productivity is high and, by so doing, tend to further raise productivity, creating an uneven distribution of activity and spatial income disparities. [1]

The aggregate impact of these locally rooted, "clusterized" innovations is to push at the far reaches of a country's aggregate technical capacity and, by so doing, compel a robust and long-range economic surge.

MULTIPLE ENVIRONMENTS, THE SEAMLESS WEB, AND HIGH-TECHNOLOGY CLUSTERS

A fundamental issue underscored by advanced materials and their linkage to competitiveness concerns what we have described as the multiple-environment context. Global business today has grown ever more complex and difficult to negotiate. A multinational company must be able to manage different aspects of its operations simultaneously. It no longer suffices that management operates exclusively within an "economically driven view of corporate strategy" [2]. Technical, scientific, financial, and political contexts must also be considered in the strategic equation. Moreover, an "intrinsic intimacy" must exist between these realms if companies, nations, and regions are to remain competitive globally. This ability to successfully operate in multiple environments is the key to securing competitive advantage in the twenty-first century. Management scholars maintain that in formulating a coherent, integrated competitive strategy the following must apply [3]:

1. Strategies pursued in each environment need to be customized to that environment's specific requirements.
2. At the same time, the strategy appropriate to one environment should not contradict or otherwise threaten other strategies appropriate in other environments.
3. There must be a close, intimate, and regular dialogue between these various strategies; the strategies of the different environments must adjust to one another as required.

These requirements find resolution in what we term the seamless web, another term for which is the industrial cluster.

We have seen that the structural context within which modern technologies in general, and advanced materials in particular, develop has shifted dramatically over the last few decades. The reign of the large corporation and its internal, self-contained R&D departments gave way to the small firm as the center of innovation, and the economic growth that attends it, within society. The implications of this shift are most fundamental. The paradigm for technological change has moved from a top-down to bottom-up model. In the former case, the most important types of technological growth occurred in isolated, self-sufficient pockets of society; specifically, certain corporate entities—the large, integrated chemical and petrochemical companies with the requisite economies of scale capable of innovating on a grand scale in an effective and economic manner. These industrial leviathans operated more or less independently of other actors making up a society's technological network in the sense that they typically funded their own R&D activities, using, for instance, retained earnings, and created their own knowledge base out of which they fashioned their new materials.

The new order of technology creation in the twenty-first century is quite different. Now, we see innovation occurring not in particular, discrete, and self-contained corporate units of minimum economic size but in numerous and proliferating smaller firms feeding off the initial research undertaken by, and thereby linking closely to, other actors making up and continuously distributed throughout society's technological community, including universities, federal laboratories, incubators, governments (in particular, state, regional, and local), and most critically, the venture capitalists. This means that, to a greater degree than in the past, technological advance may occur at any point throughout society and involve, to a larger extent than before, a far wider variety of participants, operating below the large corporation–Federal government levels, within the creation network.

Certainly, the integrated corporation itself has a role to play in this system. Although they do not wield the same measure of control as previously, they do absorb and process the new ideas and technologies created by the more numerous and widely distributed smaller firms, allied organizations, and subfederal political governments. More than at any other time in the past, we can say that new technology creation takes place within a seamless web environment in which occurs a virtual continuum of creation, transfer, and distribution through society.

This "seamless web" model in no way implies that technology creation is a chaotic and unpredictable process. Technology creation happens, in fact, in a very organized manner within a growing number of coherently formed communities or complexes. These clusters congeal in order to provide technology creation with the requisite synergies for economies of creation necessary for national competitive advantage. A fundamental dynamic of these technology clusters is that they maximize the ability of firms to operate in a concentrated multienvironment context that has become necessary in a modern, decentralized innovation process.

In general, these multienvironment clusters operate under rational and predictable laws. These clusters impose order and a harmonized interactive system on what could otherwise be a research and development activity that is too dispersed, inchoate, and ultimately inefficient to sustain national competitive advantage. The locational patterns of these clusters and their individual growth patterns and evolutionary course may be readily forecast and tracked. Each cluster is itself a seamless network of technology actors. These expanding clusters, through information, technology, and personnel transfer, link up with contiguous clusters in an area to form "mega" clusters, and extend the seamless, organized, and efficient web of creation throughout regions and states. We saw this process operate as local webs linked up to regional networks, then coordinated together in state-wide advanced-material (and nanotech) initiatives. The European mode of technology creation "from above" has not proven as effective. Its bias toward theoretical models and centralized research—what might be termed the academic–government complex—grossly underestimates the importance of the technology transfer process and the intricate and ongoing dialogue that takes place between technology, capital, and markets that makes technology creation, transfer, and diffusion possible. If all politics is local, so are technology creation and transfer, and their progeny, productivity growth and economic expansion. Europe's difficulty in grasping, or at least accepting and implementing, this paradigm, has severely hindered it from competing globally with the United States. As we have seen, structural and sociocultural factors come into play here. An aristocratic tradition that esteems the pure sciences over the "crafts" and the absence of an American-style frontier experience and an ultra- practical, "succeed in any way you can" mentality may be at work here.

Whether this is so, or whether other, more recent forces need attending to as well, it is clear that the European style of innovation stresses those elements that get in the way of the creation of a complex, textured, interconnected, and bottom-to-top web of creation, as embodied in coherent and expanding high-technology clusters. Even the most advanced of Europe's cluster creations, such as at Oxford in the United Kingdom, cannot even begin to compete on a technical or economic level with its most illustrious counterparts in the United States. It is not in the least coincidental that the very organs and sinews of a viable, flexible, and effective network that are so clearly at work in U.S. cluster arenas, appear attenuated and even dysfunctional within the European context. Given Europe's dependence on the sciences, institutional capital, and big government in the innovation process, it is not surprising that the less grand, more prosaic actors—venture and angel investor groups, incubators, local governments, and local and regional gatekeepers—find themselves out of the spotlight and marginalized in the one activity in which they must play so pivotal a role. Without them, large gaps appear in what needs to be a system that is integrated and seamless. Without a viable venture capital community, selection of appropriate technologies from the mass of candidates breaks down. Ineffective incubators cut to the very heart of technology transfer from universities to outside markets. The absence of a force of regional and local gatekeepers means that there are no galvanizing forces or centers of crystallization that can bring together the disparate actors and elements into a unified and potent cluster organization. Agglomeration economies fail to take hold. The network does not link together and, in effect, short-circuits, causing dissipation of creative energy and, in some cases, system collapse.

OPPORTUNITIES AND THREATS: FUTURE CHOICES FOR THE UNITED STATES, EUROPE, AND ASIA

What, then, does the future hold? Can we assume that this divergence, this great divide, will continue or will globalization finally begin to take effect and bring the two great economies onto a synchronous paths? There is some evidence that convergence might be in the offing. The United States faces a host of difficulties that may reduce its innovative hegemony and jeopardize its economic position in the world. These include an expanding fiscal deficit, increasing public debt levels, and a sharply weakening dollar. At the time of this writing, the United States was facing a mortgage and credit crisis of major proportions and appeared heading into an economic slowdown and possible recession. This basket of macroeconomic troubles could mean less money for the states and localities, the very source of creative clustering. Shortages of American-born engineers and the restrictive immigration policies that followed in the wake of 9/11 mean fewer trained personnel needed for technological growth, including in the advanced materials field.

It is this human-resource issue, in particular, that appears to be at the root of the growing competitiveness of the Scandinavian countries. Sweden, for example, has been very successful in attracting, retaining, and developing creative talent. In stark contrast, U.S. policy that exacerbates the "tolerance factor" seems to be limiting the entrance of foreign engineers, technicians, and scientists into the country. This means that "the United States may well be losing its long-established edge [relative to certain European countries] in attracting the brightest and best talent from around the world" [4].

If the United States weakens technologically, the forces of globalization could then place Europe in a more favorable position. The growing number of joint partnerships between U.S. and E.U. firms could mean a transfer of knowledge of skills, information, and technology from American to European companies. It is possible that we are seeing this occur already. There is evidence that the American system of technology creation and application, hinged as it is on multidimensional gatekeepers, may be filtering into Europe. If so, the extent, degree, or speed with which this is happening is not clear. Nevertheless, the case of England's Oxonica offers the intriguing possibility that change is in the wind within the European Union and in a direction that may prove a future challenge to U.S. technological supremacy. From this perspective, we may view Europe's ultimate ability to gain competitive advantage over the United States as an internal conflict between its traditional biases and predilections, such as big science over engineering practice, and the impact of globalization that compels, or at least entices, European firms and governments into adopting the American system of technology creation. We know from numerous reports, articles, and policy documents generated within the EU over the last few years, many of which were cited in this book, that, at the very least, European officials, executives, and scholars are very aware of the problems they face in technology creation vis-à-vis the United States. Whether these insights that force the United Kingdom, Germany, and France to pay homage to American ingenuity and entrepreneurial skills will (or even can) be transferred and adapted within the European context remains to be seen.

Beyond the question of Europe looms the shadow cast by the Asian economies. Are they prepared to compete with the West on equal terms? Is globalization raising them up in a process of convergence of unprecedented proportions? And what does our journey into the inner workings of the global advanced-materials industry tell us about the competitive prospects of Asia relative to Europe and the United States? China and India grab most of the attention when discussions of Asia and its putative economic prowess enter into discussions. China especially is generally perceived as the "can't miss" market for current or future international firms. Before we accept this as a truism, we should recall that in the 1980s the industrialized world perceived Japan in such terms; and Japan has yet to live up to this assumed potential.

So, where do India and China, and Japan for that matter, stand competitively in the first decade of the twenty-first century? Japan remains a highly competitive economy. India and China now rank within the world's 20 largest economies. Indeed, China occupies fourth position after the United States, Japan, and Germany. These three most important Asian economies have much in common with regard to technology creation in general, and advanced materials in particular. All three countries provide advanced technical training for scientists and engineers. Even so, they send their students to the United States for their graduate engineering studies. Increasingly, these students return to their countries to apply what they learn. This is evident in the expanding pool of patent registrations. They are spending increasing amounts of money, public and private, on new-materials research and development with the aim of linking this activity to the important consuming sectors, especially IT, communications, and biotechnology. These countries have already entered into nanomaterial research, development, and commercialization. Japan certainly leads in patenting and applications of these materials, but China and India continue to pursue this line of research at a quickening pace. Japanese and Chinese companies enter into joint partnership arrangements with western, especially U.S., R&D firms in the hopes of securing advanced-materials technology for home markets. In all three countries, the rate of advanced-materials technology transfer from university to start-up firm is on the rise.

But can we say that the major Asian economies are converging toward the American model of high-technology creation, or do they reflect the European model? In fact, there is reason to believe that they might be veering toward American-style clustering. If so, this might indicate a promising approach toward sustained economic growth. The central issue is the degree of urbanization that has been taking place in Asian society.

Economic geographers view urbanization as important for clustering, productivity, and economic growth. In the last two centuries, cities have consistently provided the environment for institutional and technological innovation, and have been referred to as engines of economic growth, agents of change, and incubators of innovation. Anthony Venables reports that recent studies that canvassed a wide range of city sizes found that doubling city size increased overall productivity by as much as 8%. This means that increasing the population of a city from 50,000 to 5 million results in productivity growth of 50% [5].

We recall that within the United States, the major advanced material technology centers are located within urban areas. There is now growing trend in the rate of growth of urbanization away from North America and Europe toward the developing countries, and especially Asia. During the period 1950 to 2000, the growth rate of urban population in Europe and North America was 1.5%. The share of Europe and North America in the global urban population declined from 53% in 1950 to 27.5% in 2000 and is expected to contract further to about 17% in 2030. It is in Asia today that almost half of the urban population lives. And soon it will have the majority of the world's urban population. Currently, six out of 10 countries with the largest urban populations are in Asia. A phenomenon in Asia is the emergence of mega-cities—large multinuclear urban agglomerations of more than 10 million people. There were no such agglomerations in Asia in 1950 and there were two in 1975. By 2000, 10 of the 17 global megacities were in Asia. By 2015, it is expected that 12 out of the 21 world's megacities will be in Asia [6] (see Table 11.1).

Table 11.1. Urban population growth across the globe (% of population)

	1950	2000	2030
North America	63.9	77.4	84.5
Europe	52.4	73.4	80.5
Asia	17.4	37.5	54.1
Global	29.8	47.2	60.2

Source: [7].

Indeed, there is some evidence of the early signs of advanced seamless web development that engenders new technology. In China, high-technology incubators are on the rise, as are such important indicators of incubator activity and health as number of tenant companies, average number of tenants per incubator, total employees in, and average income of tenant companies. There are also a growing number of home-grown venture capital firms that are entering advanced materials and nanotechnology. Between 1999 and 2002, $900 million of Chinese venture capital entered into nanotechnology companies, with 43% of this accounted for in 2002 alone. It is not surprising to find that in 2005 China ranked third in the world in terms of creation of nanotechnology patents. As Chinese companies continue to strike R&D and joint venture deals with U.S. high-technology firms, they will continue to learn from and adapt the American model of advanced-technology creation [8].

India also appears poised for expanded high-technology activity. In 1981, 23% of India's population was urbanized; in 2001, this percentage increased to 29%. Furthermore, those Indian states that are the most urbanized (Tamil Nadu, Maharashtra, and Karnataka) have recorded higher growth rates and agglomeration of industries and services, including those in the high-tech area. Thus, the southern Indian cities of Bangladore, Hyderabad, Chennai, Mumbai, and Pune have emerged as competitive advanced-technology hubs. These new thriving cities of India are all centers of excellence in terms of education, particularly technical education, technical training colleges, R&D establishments (both public and private), and high-tech industrial and service activities. These are becoming centers of such advanced technologies as IT and the critical materials needed to feed this sector [9].

Does this all mean that Asia will soon become the technological rival of the United States in advanced materials and the industries that consume them, even more than Europe? Before we can answer this, we must consider the problems that still face Asia that can severely interrupt the formation of new technological clusters. In the first place, centralization still has a tight grip over India as well as China. Privatization has proceeded exceedingly slowly. This means that both countries naturally tend toward the "top-down" model that, as we have seen, limits European technology creation and economic growth.

Then too, both China and India must still contend with widespread poverty. India's per-capita income, for instance, is just over $600. China's troubles are also severe. It is estimated that it will take a decade of growth of over 8% per annum for at least 10 years to begin to make a dent in reducing overall poverty in China. Despite a rising middle class population as a result of globalization, municipalities must, in the face of persistent poverty levels, continue to struggle with a severely constricted tax base. In Europe and the United States, municipalities receive about $2900 per capita in revenue per year, compared to only an average $150 in Asia [10]. This means fewer resources to attract the most advanced companies to nascent clusters. It also means difficulties in maintaining essential infrastructure, which hinders municipalities in India and China in undertaking new projects. Furthermore, local governments do not have the same power and prestige as in the United States, and public officials at the local level in India and China are typically at the lowest rank in public service in terms of salary structure and competence. But as we know, it is at the local level that new technology emerges, and these governments (in the United States at least) are often vital for this development. This problem reinforces the grip of the central authorities over the technology creation process and creates gaps in establishing and extending communications between organizations within and between cluster communities. Ultimately, it severely limits local policy makers from spearheading industrial growth, such as we saw in upper New York State, and which we know is fundamental to technical progress.

Further questions arise when we examine the role of private capital in Asian economies. The growing linkage of venture capital, incubators, and start-ups in India and China does not assure that the American model of technology growth has yet taken hold in Asia. Financial and capital markets still remain undeveloped, certainly by western standards. Venture and, even more, angel capital, which are such critical links in cluster communities, have not proceeded past the early stages of development. A far greater capability than currently exists must be reached for potential lenders to appraise technology projects as reasonable risks.

Beyond the ability to judge risk, the need for more risk taking on the part of venture capitalists, incubators, and start-ups must become part of the equation in these countries. This means more privatization of these sectors. The government still owns and operates most of the incubators and, either directly or indirectly, the bulk of venture capital funds in China. This means that cautious bureaucrats with limited experience and vision, rather than entrepreneur–gatekeepers, control these most important centers for new technology. These risk-averse, government-run incubators take little or no equity stake in the companies they foster, a system that severely curtails any incentive to help tenants succeed. By the same token, venture capital filters into those projects with the greatest chance of success (i.e., least risky) because companies that accept such funding from the government are careful, for political as well as economic reasons, not to lose public funds. Clearly, this mind-set significantly reduces the chances that firms will pursue the most important, and generally high-risk, projects [11].

Ultimately, it is unclear whether either India or China will embrace a culture from which the all-important gatekeeper can emerge. Asian universities tend to place great stress on theoretical construct. Engineering science rather than engineering practice continues to dominate. There are precious few programs in which engineering students spend part of their studies within an industrial context. Thus, they obtain very little real-world experience in the commercialization process. With increasing specialization the norm, students do not know how to coordinate technology, markets, financial analysis, and venture capital formation in the joint service of innovation.

Although there are strong indicators of technological growth in Asia, such as the rise of incubators and other venues of technology transfer, neither India nor China have as yet demonstrated the ability to create world-class industrial clusters. So, though there are positive signs in both the Europe and Asia for future competitive leadership, there are also red flags that point to potential landmines that can hinder, and even scuttle, these emerging countries' push to be competitive powers on the world stage. The American model of technology creation and international competitiveness remains in force as the gold standard for progress. Whether the United States can retain its enviable position is a reasonable and pressing question as the world proceeds to globalize. If the United States loses its place, will there be an advancing convergence of economic power, or will divergence remain in force, but with a shift of leadership to another country or region of the world? Finally, what does this understanding of the convergence versus divergence issue tell us about global economic development generally? Certainly, the poorer countries in Eastern Europe, Asia, Africa, South America, and the Middle East are not yet in a position to join competitive battle in the type of high-technology environment within which the United States and the European Union now operate and contend. However, given the results that we have found in our study, is it not a fair question to ask whether development of home-grown small and medium sized enterprises interlinked in a bottom-to-top "seamless web" structure composed of a network of local entrepreneurs, venture capital, forward-looking local political initiative, and even the beginnings of incubators, in such areas of production as apparel and clothing, metal fabrication, electronic assembly, and other "low-tech" but economically crucial industries can serve as the early foundation and necessary precondition for today's less-developed countries to evolve over time into more technologically advanced economies, thus securing internally generated, and thus self-sustaining economic growth over time? If so, can we not say that this is a true path to convergence through globalization, a path that seriously questions, as William Easterly [12] reminds us, the ". . . Depression's deceptive intellectual legacy . . . that development [of the poorer nations by the West] flows from [above to below] by all-knowing states rather than [from below to above by highly localized] creative individuals"? Then, the answers to these questions must surely be found by closely following the temporal and geographical patterns of the world's exceptional twenty-first century technologies, and most particularly, the advanced materials industry, which, the occasional recession notwithstanding, is the bellwether and indeed fundamental impetus of modern nations' deep-rooted and long-term economic fortunes.

REFERENCES

1. Venables, A. J. (2006), "Shifts in Economic Geography and Their Causes," *Federal Reserve Bank of Kansas City Economic Review,* Q. IV, p. 19.
2. Cummings, J. and Doh, J. (2000), "Identifying Who Matters: Mapping Key Players in Multiple Environments," *California Management Review,* Vol. 42, Issue 2, p. 83. For the role of gatekeepers operating in multiple environments, see Gittelman, M. and Kogut, B., "Does Good Science Lead to Valuable Knowledge? Biotechnology Firms and the Evolutionary Logic of Citation Patents," *Management Science,* Vol. 49, No. 4, pp. 366–382.
3. See for example Baron, D. (1995), "Integrated Strategy: Market and Nonmarket Components," *California Management Review* Vol. 37, Issue 2, pp. 47–65.
4. EurActiv.com (2005), "Sweden Beats US as Most Creative Nation," *EU News and Policy Positions: Innovation and Jobs,* November 22.

5. Venables, A. J. (2006), "Shifts in Economic Geography and Their Causes," *Federal Reserve Bank of Kansas City Economic Review,* Q. IV, p. 25.
6. Mohan, R. and Dasgupta, S. (2003), "The Twenty First Century: Asia Becomes Urban," Urban Symposium Keynote Address, World Bank, Washington, DC, December 15, pp. 6–10.
7. Ibid., p. 2.
8. Kanellos, M. (2003), "Nanotech Spending Nears $3 billion," CNET News.com, June 25; *People's Daily Online* (http://english.peopledaily.com) (2003), "China's Nanotech Patent Applications Rank World's Third," October 3.
9. Mohan, R. and Dasgupta, S. (2003), The Twenty First Century: Asia Becomes Urban," Urban Symposium Keynote Address, World Bank, Washington, DC, December 15, pp. 21–22.
10. Ibid., pp. 10–11.
11. Harwit, E. (2002), "China's High-Technology Incubators: Fuel for New Entrepreneurship?" East-West Center, Economics Series No. 1 (Honolulu, Hawaii), pp. 3–4, 8–10.
12. Easterly, W. (2008), "Development Doesn't Require Big Government," *Wall Street Journal,* October 3, p. A19.

Index